Penguin Education

Management of Production

Edited by M. K. Starr

Penguin Modern Management Readings

General Editor

D. S. Pugh

Management of Production

Selected Readings

Edited by M. K. Starr

Penguin Books

Penguin Books Ltd, Harmondsworth,
Middlesex, England
Penguin Books Inc., 7110 Ambassador Road,
Baltimore, Maryland 21207, U.S.A.
Penguin Books Australia Ltd, Ringwood,
Victoria, Australia

First published 1970

Made and printed in Great Britain by
Richard Clay (The Chaucer Press) Ltd, Bungay, Suffolk
Set in Monotype Times

To David B. Hertz
Whose vision has helped keep the management sciences vital

Contents

Introduction

This collection of production readings has several strong, thematic threads that run through it. However, traditional threads and their fabrics are usually the most apparent. Therefore, it is my intention to point out beforehand a primary organizing principle that was used, which is less visible than the traditional components. Like the 'emperor's clothes', things may not seem to be what others say they see. The reader will judge for himself.

It is a major reflection of the wiser men of our era that instead of looking for the right questions to ask, so many are looking for answers to their problems. From systems concepts we know that often the tragic part of trying to go straight to the root of the problem, while seeming logical, is that the effort involved becomes part of it, changing the original problem.

We teach production management, or operations management, or whatever it is called at a particular time and given place, in as straight a line as we can find. We logically classify problems and go straight to their respective roots, thereby creating a new class of systems problems. Production management, no matter by what name it goes, will continue to be taught in this manner as long as production is thought of as a unique and segmented line function. If production management is responsible only for measurable output achievement, and not for over-all company achievement, the straight line concept will persist. I have not intended to make a pun about the 'line function' and taking a 'straight line' to it. The appearance of linearity in both contexts is not coincidental.

The table of contents of this book is divided into sections that represent a logical (straight) line of thought. It begins in Part One with product and project development. The planning process rapidly interrelates with process design (Part Two). Policies for operating the system are dependent upon the product and process decisions (Part Three). Consequently, Part Four suggests that a larger systems view can emerge which will help to relate the three previous parts to each other. In so doing, this table of contents is up to date, emphasizing both analysis and synthesis in the modern systems sense.

To pursue the 'other-than-line view' requires cultural breadth. This quality of breadth has not been part of the production managers' job description. That it should be is the view that we will take. The cultural sense is epitomized by *generalization* rather than *specialization*. The latter is characterized by a narrow field explored in depth. The total scope of such a system is the aggregate of a set of narrow fields. The rules for interconnexions are tenuous. In the production field, specialization of this sort has been more evident than in other functional areas. Problems concerned with inventory, time studies, production scheduling, quality control, etc., are analysed individually and thoroughly. The aggregate of many such specializations, taken in this way, is not what we mean by generalization. It is not even a reasonable approximation to it.

Studies that range across many different topics where several areas are simultaneously interrelated begin to approach the kind of generality we have in mind. Accordingly, within each section of this book's table of contents, there is a high variety level of topics and techniques, and a large variance of opinions and conclusions. The articles have been selected with the intention of lighting up a broad horizon of interconnexions rather than pin-pointing with intensity special areas of interest. The justification for this approach deserves to be expanded upon.

At the turn of this century, the notion of culturally broad production managers would have seemed peculiar and to many entrepreneurs, undesirable. But this is no longer so and it was not true long ago. To illustrate the 'long ago' first, one of the most creative production men of all time was Benvenuto Cellini (1500–71). His production efforts (in this craft era) were limited to one, or a few items, produced at a time, such as the famous salt-cellar of Francis I. The eighteenth-century Industrial Revolution introduced a different kind of creative potential which attracted contributions from the most innovative minds of that period. Production, as epitomized by Eli Whitney and Leblanc, was totally creative in the conceptions of how the production system could be managed.

Once the system was working, grinding out increasing volumes of production, the issues shifted. Efficiency was critical. To cut down the cost on highly repetitive operations meant that steps needed to be taken which fell outside the realm of man's ordinary scale of response and sensitivity. A hairbreadth of a motion, an extra couple of grains of material could sum over many repetitions to significant quantities. This was hardly an enlarging view of the industrial world for the production manager. He turned inward and became a microscope for detecting seemingly trivial flaws that he knew were not trivial because of the scale of production.

Engineering improvements paced these developments and everyone had just about the same idea of what was the right thing to do. Henry Ford's invention of sequenced assembly and his management of this concept, while totally creative in their own right, brought further rigidity to the situation. All such developments fostered efficiency and reinforced specialization. It was hard to be creative when the handbook told the story like a grown-up version of a child's construction toy: step three follows two follows one, etc. One should not confuse such logic with simplicity. Production has never been simple; not before nor with Cellini, not with Taylor nor with Ford. Production abounds in detail, and its management promptly envelops anyone who enters the field, leaving little room for developing breadth, imagination or creativity.

The environment for production was stable only as long as the output objectives of the system remained unchanged. Now, however, these objectives of production management are undergoing marked change. Forecasting has new importance to production management because the objective of volume at the lowest cost is no longer the main (or only) concern of management. Consumers have become unreliable. They have ceased buying *just* the benefits of mass production. And as forecasting became essential, small systems considerations began combining to form larger systems for which meaningful forecasts could be developed. Single product systems started to disappear in the face of larger product-line advantages.

Today's study of production management must be devoted to the significance of such growth. With small systems, operating in a free, competitive market-place, the production system did not outgrow the market. On the contrary, the problem was one of obtaining funds for additional capacity because output goods were scarce and the market could absorb all that could be produced at a reasonable price. Economic reversals in this pattern did occur, producing a general decrease in buying power. But such recessions and depressions create conditions that are dissimilar to those caused by the demands of affluent, restless consumers. Depression is a cutback phenomena, that exerts more pressure on the production system for efficiency. The present kind of change, however, exerts major pressure on the production system to enlarge its output, to evolve its product-line; in sum – to produce a broader spectrum of items from which the consumer can choose.

Production management *can be* crucially helpful in finding innovative alternative configurations to satisfy these demands, at a reasonable cost. The management of *new* (and available) technology is at the heart of this issue. The technology exists but it is not being utilized. The consumer seems to know this. Maximizing the consumer's choice is not a force exercized by industry upon the consumer. The reverse is true. The consumer is forcing upon industry the demand for enlargement of his selection base. One way to be relatively sure that this position is correct is to note that industry has been resisting these developments for change. A variety of sources close to industry support this contention. Some of the explanations for the inertia might be the following:

1. Top management knows the production function least well.
2. Production capabilities have changed more fundamentally in the last two decades than any other functional area of industry.
3. The understandable inertia of successful, operating systems prevents *major* redirection of funds to new opportunities and greater risks.
4. Since major investments are required to move in the direction of new production systems that can help to maximize consumer choice, even small risks must be applied to larger investment bases.
5. Tactics for delaying the forces for change are available

through marketing manoeuvres, but increasingly, they are less successful.

In spite of delays, consumers are becoming relevant variables in the equations of the production process. Consumers are forcing production to innovate, if indeed innovation is the making of changes in previously accepted practice. As a result, the importance of the production management role in the world of business is increasing. It is no longer possible (as it was for Frederick W. Taylor and the rationalist school of production management) to say what is the 'one best way' for production to follow. Choice exists and ingenuity pays off. Companies will not imitate each other. This is true even though production remains an almost totally visible function.

In the previous period, when 'good' production practice was remarkably similar within industries, one could hardly claim that espionage was responsible for the rapid acceptance and spread of 'how it should be done'. Unlike the areas of marketing and finance, measurable transformation is at the heart of the production system. Whether these transformations are mechanical, electronic or chemical; whether they are spatial (as in travelling from one point to another) or informational (as in reduction of a set of data to an average) the production function is visible, but more important, *rational.* Under the new pressures and opportunities, the management configuration becomes so complex and so individual that each organization has its own best way. As soon as production is viewed as part of a larger system within its organization, its behaviours no longer isolated, it becomes representative of its company and not of handbooks that are widely published and distributed. It becomes more a part of particular management and less a part of universal engineering. To successfully achieve what is required production managers sorely need the cultural breadth that characterizes larger system participants.

There are many conceivable directions, and what is 'best' can only be judged when it is aligned with a total system design. Early production management accents on efficiency were based on stable configurations of small systems. But these assumptions do not apply to *effectiveness*. The creative potential of efficiency objectives is very small when compared with the potential of effectiveness. There is no point in making the best brass knob in the business if a plastic one will soon replace it; no point in reducing the cost of the buggy whip handle when the gear shift is part of tomorrow.

To be *creatively effective*, production management must participate in *planning* and major policy formulation. Otherwise, it will be restricted to the limited role of an enforcement agent of policy and occasionally as an agent of policy interpretation. Only through participation in the planning function can it deal with alternatives for major investment. That is where it is now capable of making important contributions to the larger system of which it is a part. Considering changes in technology, methodology and the impact of computer systems, surely the difference between old-time and modern production stands out.

The change is massive. The potential impact of new technology is just part of the story. Technological change must be considered to go hand in hand with methodological developments and computer support. Only in combination can the new technology emerge. Economic sensibility is essential to evaluate the new products and processes and to position them within the organization. At the same time, other systems factors appear, forcing a broader systems view of production management's role. Society demands pollution controls and high-level safeguards from faulty product design. Satisfying the functional needs of the consumer is only part of the production manager's job. Social expectations expand and become greater each day. A deeper understanding of the meaning of production management is essential to deal with this change. Production and distribution management interrelate. Marketing claims, product qualities and guarantees merge in congresses, parliaments and the courts. The nature of a reasonable price, always a competitive issue, now becomes a social one too.

As production management moves up in the organizational hierarchy it does so by recognizing that synthesis is essential and questioning how it can be achieved. Starting with effectiveness, we move to efficiency; starting with systems planning, we move to operations. Such cycles, which capture the flavour of the large and small systems, continue to operate until temporary satisfaction is obtained; then they start again. To many, such dynamism causes insecurity. Only to the culturally broad individual is it a realistic challenge reflecting the broader aspects of our present view of life.

The compiler of a book of readings surely comes to realize how grateful he is to the many outstanding contributors from whose materials only a small selection can be made. It is a pleasure to carry the burden of such an excess of riches. But to pay for such pleasure there is a larger task to contend with than one might initially suppose. Good ideas for organizing the book help immensely if backed up by administrative procedures for tracking the literature, corresponding with authors, assembling materials and communicating with the editors.

In each of these phases there are acknowledgements and thanks to be given, but let me begin by expressing my gratitude to all of the authors who are represented. Many of them made special efforts to bring their writings to the most useful and current stage of development. Dr Peter V. Norden worked at double time to completely update his article in line with our schedule requirements. I appreciate the significant administrative assistance received from Mrs Judith Dumas whose efforts enabled the flow of materials across the Atlantic to always be smooth. Thanks are due to Messrs James P. Monahan and Alan C. Renda for their careful and insightful help in obtaining references and checking galley proof and pages. And lastly, let me point out that the Penguin people are fun to work with.

Part One **Product and Project Development**

To begin with . . . there is the product *idea*. Idea generation is not easily assigned and is everyone's concern. After the idea . . . the *feasibility* study follows. Feasibility determination rests firmly in the production manager's bailiwick, but it is not accomplished by just technique and technology. Feasibility is a big concept requiring *co-ordination* of marketing, finance and production information. Because technology is highly specialized, and because *now* it offers so many degrees of freedom, production management is the logical candidate for the co-ordinator. Logical or not, co-ordinating multi-functional systems is not one of production management's present-day strengths. This important inconsistency must be overcome.

A feasibility study can *incorrectly* designate success (Type two error) or *incorrectly* signal failure (Type one error). The probability of any errors is inversely proportional to the degree of co-ordination achieved. Which type of error is worse? Consider the costs of a Type one error. First, there is the cost of the rejected study. Then there is the cost of not having studied a product or process idea that could have been a winner, not a loser. The latter is usually far the largest system's cost, but, since it is hidden, there is an *understandable* tendency not to worry about it. What are the costs of a Type two error? The most severe situation occurs, if the acceptance is a mistake. Major investments of time and money go down the drain, and so does morale. However, *even when* profit is obtained, the cost of not having found the most profitable idea cannot be ignored. The difference between the best and our choice is not likely to be great when a consistent search policy for new ideas has prevailed. This cost-creative orientation is the backdrop to Part One.

1 M. Asimow

The Morphology of Design *and* The Feasibility Study

Excerpt from M. Asimow, *Introduction to Design*, Prentice-Hall, 1962, pp. 11–23.

The Morphology of Design

The Design Project

Each design project has an individual history which is peculiarly its own. Nonetheless, as a project is initiated and developed, a sequence of events unfolds in a chronological order forming a pattern which, by and large, is common to all projects. We wish to examine this pattern, and to bring into sight the methodology of design by which ideas about needs are projected creatively into ideas about things; and which in turn are translated into engineering prescriptions for transforming suitable resources into useful, physical objects.

First we consider the over-all pattern. A project goes through a series of major phases. Generally, a new phase is not begun until the preceding one has been completed, although sometimes final details have to be attended to while the next phase is in progress. Often, particularly when the project is large, much of the personnel will change with each new phase in order to bring to bear special skills and knowledge, and only a skeleton group, a project team, will remain permanently with the project. We will enumerate and describe these phases, sketching in briefly the steps of which they are composed. The morphology of design refers to the study of the chronological structure of design projects. It is defined by the phases (Figure 1) and their constituent steps. The steps of the design phases, having been put into perspective, can be enlarged.

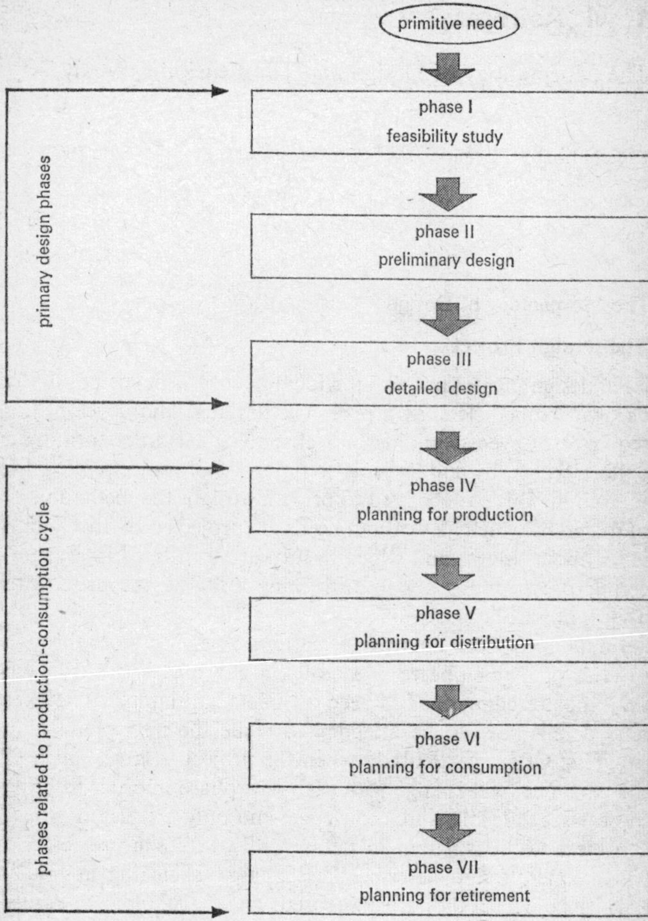

Figure 1 The phases of a complete project

Feasibility Study – Phase 1

A design project begins with a feasibility study; the purpose is to achieve a set of useful solutions to the design problem. Sometimes a design group is assigned to a project for which a design concept has already been fixed. This implies one of three possibilities: a feasibility study had been previously made; the technical management has had so much experience with the particular design problem that further study should be superfluous; the management, by omitting the feasibility study, is proceeding on unsupported intuition.

The first step in the study is to demonstrate whether the original need, which was presumed to be valid, does indeed have current existence, or strong evidence of latent existence. The next step is to explore the design problem engendered by the need and to identify its elements such as parameters, constraints, and major design criteria. In the following step an effort is made to conceive a number of plausible solutions to the problem. Finally, the potentially useful solutions are sorted out of the plausible set in three steps on the bases of physical realizability, economic worthwhileness, and financial feasibility. In conclusion, the completed study indicates whether a current or a potential need exists, what the design problem is, and whether useful solutions can be found; that is, it investigates the feasibility of the proposed project.

Preliminary Design – Phase 2

The preliminary design phase starts with the set of useful solutions which were developed in the feasibility study. The purpose of preliminary design is to establish which of the proffered alternatives is the best design concept. Each of the alternative solutions is subjected to order of magnitude analyses until the evidence suggests either that the particular solution is inferior to some one of the others, or that it is superior to all of the others. The surviving solution is tentatively accepted for closer examination. Synthesis studies are initiated for establishing, to a first approximation, the fineness of the range within which the major design parameters of the system must be controlled. Further studies investigate the tolerances in the characteristics of major components and critical materials which will be required to insure mutual compatibility and proper fit into the system. Other studies examine the extent to which perturbations of environmental or internal forces will affect the stability of the system.

Next, projective-type studies are undertaken, addressed to the question of how the solution will fare in time. The socio-economic conditions, such as consumers' tastes, competitors' offerings, or availability of critical raw materials, may change; the state of technical art may advance; and eventually corrosion, fatigue, and deterioration of performance may set in. Time will almost certainly erode the quality of the product. The question is, how fast? The rate of obsolescence or wear must be accounted one of the important design considerations, and its economic impact must be put in the balance. Finally, critical aspects of the design must be put to the test in order to validate the design concept and to provide essential information for its subsequent phases.

Detailed Design – Phase 3

The detailed design phase begins with the concept evolved in the preliminary design. Its purpose is to furnish the engineering description of a tested and producible design. Up to this point the design project was characterized by great fluidity. Major changes in concept could be accommodated without great financial loss. Indeed, for the first two phases such fluidity is essential, for they are primarily exploratory in nature, seeking to reveal an adequate range of possible solutions. At this point, however, either exploration on a large scale must come to a close, and a final decision for a particular design concept be made, or the project must be abandoned as infeasible. In the latter event, final abandonment, if warranted by favorable circumstances, may be deferred pending the results of an additional search for possible new solutions.

With the design concept in mind and the preliminary synthesis information at hand, an over-all, but provisional, synthesis is accomplished. It is developed as a master layout. With this as a basis the detailed design or specification of components is carried forward. From time to time, exigencies in the detailed work at the component level may dictate changes in the master layout; therefore it has provisional status. As the paper design progresses, experimental design is appropriately initiated. Experimental models are constructed to check out untried ideas which are not suitable to final disposition by analysis. Components, partial prototypes, and finally complete prototypes are tested as the need for information arises. This information, accruing from the testing programs, provides a basis for redesign and refinement until an engineering description of a proven design is accomplished.

Planning the Production Process – Phase 4

Whereas the preceding three phases were peculiarly in the province of the engineering designer, much of the responsibility for phase 4 will be shared with other segments of management. A new battery of skills, those of tool design and production engineering, come into play; the original project group, however, may continue in its role of leadership. The decision to produce often involves an enormous economic commitment. The level of confidence in the success of the product must be very high to support a positive decision. The decision itself must be made at that level of management at which the final responsibility for the success of the enterprise rests. The evidence on which the engineer responsible for the design project bases his confidence must be communicated in a condensed, but nevertheless fully revealing, form to the appropriate decision-maker. Ideally, the designer's confidence will be shared by the superior, who will re-evaluate this confidence using additional information concerning financial capability, business conditions, and like considerations, before rendering a final decision.

The production planning phase involves many steps which will vary in form and detail with the industry concerned. The following abbreviated list is typical of the mass production industries:

1. Detailed planning of the manufacturing processes as required for every part, every sub-assembly, and the final assembly. The information is usually displayed on process sheets, one (or several if space requires it) for each part or sub-assembly. The process sheet contains a sequential list of operations which must be performed to produce the part; it specifies the raw material, clarifies special instructions, and indicates tooling and machines

required. This step is particularly important, because design features that lead to difficulties in production are revealed. Such difficulties should have been minimized by previous timely consultations between product designers and tool designers. Similarly, questions about materials should have been resolved by consultation with process metallurgists.

2. Design of tools and fixtures. This design work proceeds generally from the information developed in the operations analysis on the process sheets.

3. Planning, specifying, or designing new production and plant facilities.

4. Planning the quality control system.

5. Planning for production personnel. Job specifications are developed, standard times are determined, and labor costs are estimated.

6. Planning for production control. Work schedules and inventory controls are evolved. Standard costs for labor, materials, and services, are established and integrated with the accounting system.

7. Planning the information-flow system. Information necessary for transmission of instructions and provision of feedback for control is determined. Appropriate forms and records are designed and integrated with computers when available. Flow patterns and routings are established.

8. Financial planning. Usually, large sums of money are required to initiate production of a new product. The source of the financing must be carefully established, and the means and rate of recovering the capital determined.

Planning for Distribution – Phase 5

It will be recalled that production is the first process in the production–consumption cycle of the socio-ecological system. The second is distribution. Although the engineering designer may not be directly involved in planning for distribution, he will often find that the problems of distribution have an important impact on the original design of the product. It is often primary, as in the design of a water or electrical power system.

The purpose of this phase is to plan an effective and flexible system of distribution of the designed good. The abbreviated list is indicative of the planning for distribution.

1. Designing the packaging of the product. The outer shape of the product may be influenced by the need to effect economy in shipping costs. Individual and special packaging may be needed to secure protection from shock and weather. Special strapping and palletizing may be needed to facilitate handling.

2. Planning the warehousing system. The economically favorable locations for warehouses are determined; the warehousing facilities are designed.

3. Planning the promotional activity. Technical sales brochures based on design information and test data may need to be developed.

4. Designing the product for conditions arising in distribution. Such factors as a shelf-life, attractive display, and final conditioning before delivery to the consumer may affect design of the product. There may be a need of enough flexibility in the design to allow of special modifications to suit customers' needs, or of further adding available optional features as required by the customer, or for modular additions to the system to enlarge its capacity.

Clearly, some of the factors enumerated here cannot await consideration before the effort for this phase will have been fully mounted, but must instead be anticipated in the earlier phases.

Planning for Consumption – Phase 6

Consumption is the third process in the production consumption cycle. Its influence on design is profound, for it pervades all phases. As a process it occurs naturally after distribution. As a phase, in the time-pattern of the design project, most of the process of consumption must be anticipated in the early stages of design in order to have a timely impact. Therefore, it is for the most part a diffused phase concerned with consumers' needs and utilities, and mingled with and attached to the prior phases. It is set up separately only to emphasize some of the special contributions which it makes in addition to its more general and pervading influences. This separate status as a phase does not imply that consideration even of the steps enumerated here as belonging to the consumption phase is to be deferred until this late stage; in fact, for most of the steps, the contrary is true.

The purpose of this phase is to incorporate in the design adequate service features and to provide a rational basis for product improvement and redesign. The steps follow:

1. Design for maintenance.
2. Design for reliability.
3. Design for safety.
4. Design for convenience in use (taking account of human factors).
5. Design for aesthetic features.
6. Design for operational economy.
7. Design for adequate duration of service.
8. Obtain service data that can provide a basis for product improvement, for next-generation designs, and for the design of different, but related, products.

Planning for Retirement of the Product – Phase 7

The fourth process in the production–consumption cycle is the disposal of the retired product. For large and semipermanent installations, the mere removal may pose difficult engineering problems, as, for example, the demolition of a tall building closely confined by a building on either side. Sometimes the impact on a new design is more immediate as when an old structure or system must be replaced by a new one with minimum disruption of normal operations.

What determines when an economic commodity in use, such as a consumer's product, a commercial or industrial device, or a private or public system, has reached an age at which it should be retired? This is one of the principal questions raised by a study of engineering economy. If the good in use has worn to the point at which it can no longer render adequate service, then the need of replacement is clear. However, the same fast pace of technology which presses upon the designer also accelerates the ageing process of goods in use. It is a peculiar mark of our times that goods in use are retired more frequently because of technical obsolescence than for physical deterioration. Changes in fashions, often deliberately cultivated by industry, also produce their share of casualties. In the design of soft goods, such as clothing, exploiting fashion changes is an accepted and respected practice, as the intrinsic value of such goods resides largely in their aesthetic appeal, but grave ethical questions are raised when obsolescence of style, unaccompanied by significant technical improvements, is used as a means of inducing the retirement of the products of engineering design.

To the engineering designer, the question whether to design for physical deterioration or for technical obsolescence is of fundamental importance. Ideally, the system should be designed so that it wears out physically as it becomes technically obsolescent; then no extra cost would be incurred by having provided for a longer than useful life. But usually, the elements of design that contribute to a longer life are also essential to adequate reliability and maintenance; thus a full balance between obsolescence and wear out is generally not possible. These aspects of design need further exploration.

What values are available when an economic good reaches a terminal point of service? And how do these values influence design? The latter question is the concern of the retirement phase in design. The purpose of this phase is to take account of the problems associated with retiring and disposing of a product. The same comments, made about the steps in the consumption phase reflecting their diffusion into the preceding phases, applies to most of the following steps:

1. Designing to reduce the rate of obsolescence by taking into account the anticipated effects of technical developments.
2. Designing physical life to match anticipated service life.
3. Designing for several levels of use so that, when service life at a higher level of use is terminated, the product will be adaptable to further use at a less demanding level.
4. Designing the product so that reusable materials and long-lived components can be recovered.
5. Examining and testing of service-terminated products in the laboratory to obtain useful design information.

In summary, although the first three design phases, forming a primary set, are the principal concern of the design group, the remaining four, which constitute a secondary part stemming from the production–consumption cycle, have so great an impact on the design that they must be considered in detail in the total design effort [. . .]

The Feasibility Study

We return to the primary design phases to look more closely at their constituent steps. The feasibility study with its steps and their interconnexions is presented in the flow diagram of Figure 2.

The Need – Establishing its Economic Existence

The starting point of a design project is a hypothetical need which may have been observed currently on the socio-economic scene. It may be phrased in the form of a primitive statement resting on untested observations; or it may have been elaborated into a sophisticated and authenticated statement based on market and consumer studies. The need may not yet exist, but there may be evidence that it is latent, and that it can be evoked when economic means for its satisfaction become available. The need may be suggested by a technical accomplishment that makes the means of its satisfaction possible. In whatever way the need has been perceived, its economic existence, latent or current, must be established with sufficient confidence to justify the commitment of the funds necessary to explore the feasibility of developing the means satisfying it. By economic existence of a need we shall mean that individuals, institutions or society will recognize the need and will pay the price of the means which can satisfy it. The means may be available in the market-place for a purchase price, or it may be proffered by a government agency and paid for by taxation. The importance of establishing the need can hardly be overstated. Too often an organization will plunge into a project and develop a technical success while achieving a financial failure because the assumed need was ephemeral and disappeared in the light of reality.

Intuitive knowledge about people, their habits and mores, and their behavior in the socio-economic system, may be combined with specific information obtained by market research to provide

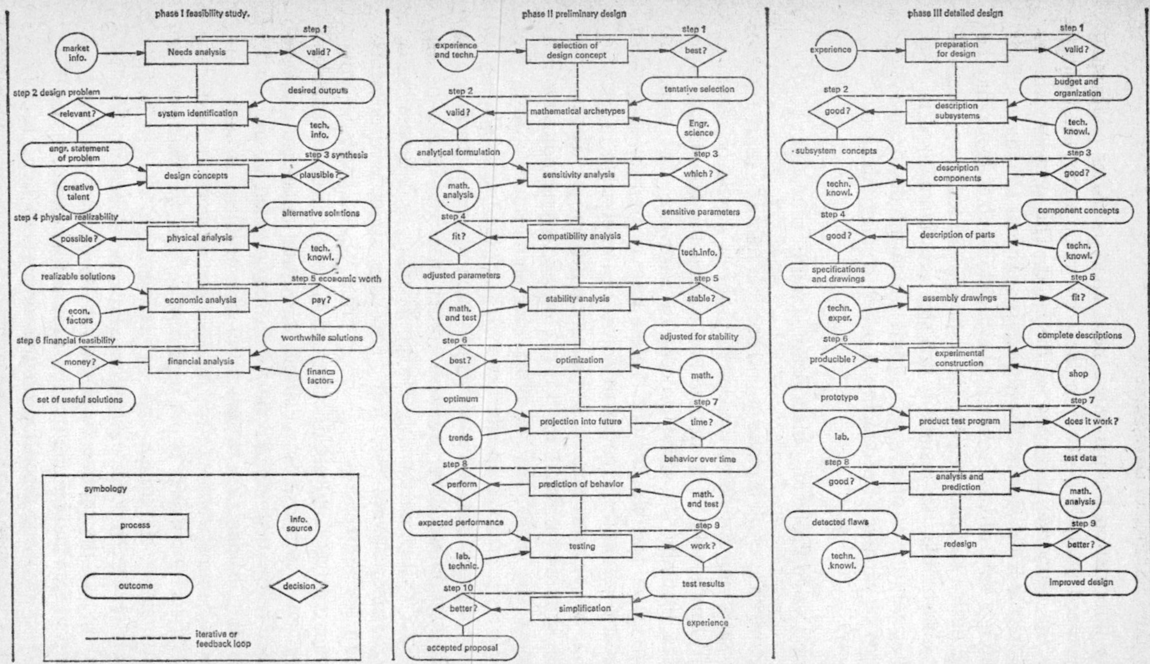

Figure 2 Morphology of design

the information for making a need analysis. After performing the need analysis a decision must be made about the validity of the economic existence of the need. The decision will be favorable if the weight of evidence is sufficient to induce a high enough level of confidence in the decision-maker (who may be the design engineer bearing the responsibility for the project). If it is favorable, the results of this step are summarized in a set of specifications of desired outputs which the device or system must be capable of producing in order to satisfy the need.

The Design Problem – Identification and Formulation

Before an attempt is made to find possible solutions for the means of satisfying the need, the design problem should be identified and formulated. It is strange how strong the temptation is to seize mentally on some concept of hardware that seems to provide a feasible solution before the real problem is well understood, and thereafter to patch up the concept in perilous ways as deficiencies in the solution begin to appear. The temptation should be resisted, for it tends to lead to a mental rut which blocks the truly creative effort that should follow after the problem is grasped. In effect, we consider the ultimate solution to be a black box, the contents of which remain unknown. The information we have available comes from the results of the preceding step, particularly the specifications of desired outputs, and from relevant technical knowledge about environments, resources and general engineering principle. With this information an activity analysis is performed whereby the design problem is given a technical formulation. The question which must be asked before this step may be considered complete is whether the resulting engineering statement of the problem is sufficiently relevant and adequate to commit the ensuing steps to the design. If the decision-maker is not confident enough, he will reject the statement and require another performance of the activity analysis step. Indeed, new information, developed in this step, may reduce the confidence in the validity of the preceding need analysis step and require further study at that level. This iterative nature, involving repetition of steps or parts of steps, is characteristic of design. Design is too complex a process to admit of an uninterrupted progression without backing up now and then to correct or rework previous results. New information is constantly developed by the design work itself which previously was either overlooked or unknown. This new information changes the confidence levels on which prior decisions were made. If the deterioration in confidence is enough to destroy the basis for a particular decision, the decision-maker will require that the affected step be reworked until the necessary level of confidence is restored.

The Synthesis of Possible Solutions

Synthesis refers to the fitting together of parts or separate concepts to produce an integrated whole. The synthesis step formally begins after the design problem is well understood, although some notions about possible solutions may have already been suggested during the prior steps. The point to be emphasized is deceptively obvious; concentration on possible solutions should not begin until the design problem has been studied and identified, and a reasonably good working formulation of the problem set down. As with prior steps, there is an iteration process, a feedback. The solutions, which are conceived and matched against the background of the problem statement, yield new insights and items of information about the preceding steps which can then be corrected if necessary.

It is the synthesis step which most characterizes the project as a design undertaking. This, more than any other step, requires inventive and creative effort. Creativity is therefore a very essential ingredient of engineering design. In the context of design we offer the following as a definition of creativity: a talent for discovering combinations of principles, materials, or components, which are especially suitable as solutions to the problem in hand. None of the individual elements which comprise the synthesis need be new or novel. Developing new and novel elements is more the object of research than of design. After a new principle is discovered by the researcher, the designer is often assigned the task of evolving some new components that are advantaged, for the first time made possible by the new principle. In doing so he will generally combine the new principle with several old ones to achieve the final result.

Physical Realizability

The outcome of the synthesis step is a set of plausible solutions. Each one of the set is a mental abstraction which takes account of some of the major factors or elements on which it depends. Omitted from such an idealization are many ancillary and consequent factors which may tip the balance towards success or failure. The problem is whether it is possible to accomplish such a practical physical embodiment as is suggested by the concept. The designer can visualize to a limited depth only the elements and results of a new concept, like the chess player who can see no more than a few moves ahead, only dimly perceiving all of the implications.

We assume that the solutions are plausible. By this we mean that the major elements of the concept could combine as visualized, in concordance with natural principles, to effect the solution; that is, in the over-all view the concept could work in principle if the supporting elements could be realized. Thus the problem of establishing, on a sufficient level of confidence, the possibility of physical realization of a solution is transferred to a similar consideration of the constituent elements. This process of transferring attention to a lower order of sub-problems is a distinctive feature of the decision process in engineering design.

Economic Worthwhileness

No object is a proper subject of engineering design if it is unable to pass the test of economic worthwhileness. Literally, this means that the object of the design must be '... of sufficient value to repay the effort'.[1] Value has a connotation which is singularly personal; it depends on the evaluator, his viewpoint, and the prevailing circumstances. The only objective measure is in the market-place. When this instrument of measure can be applied, the results can be quantitatively reckoned and expressed in terms of dollars. Otherwise values remain subjective. Indirect tools, highly sophisticated and difficult to apply, must be used for their measure.

1. See *Webster's New International Dictionary*, unabridged, 2nd edn.

It is of interest to contrast the values of producer and consumer. The producer must acquire the requisite resources of raw materials, energy, capital, and manpower. These he can obtain on the open market; therefore their value of acquisition can be determined objectively. The final product, after its completion by transforming the input resources, is returned to the market-place. Again, the value can be determined, and the input–output values compared. If the flux of values through the production process has occasioned an augmentation of value sufficient to induce its continuance, the process is deemed to be economically worthwhile. The extension to the distributor follows in the same manner.

The consumer can likewise measure the value of acquisition by the cost of the product in the market-place. With this product he can satisfy appropriate needs. But how does he, or a detached observer, measure the value of this gratification? Clearly the value is subjective and cannot be measured in the market-place; therefore the objective monetary value of the cost is not commensurable with the subjective use value. We can only infer economic worthwhileness by the continued willingness of buyers to pay the purchase price. Nevertheless, the designer must predict as best he can the intensity of the economic worthwhileness to consumers in general so that he can estimate the market potential; for his choice of a design concept must be conditioned by the economic worthwhileness of each and all of the chief protagonists in the production–consumption cycle. It is for this reason that the designer must be prepared and able to place himself mentally in the economic and psychological states of each in turn: producer, distributor, and consumer. As he visualizes himself in each role, he must extract the essential contributions to the design which each viewpoint offers. Clearly, those solutions which do not pass the test of economic worthwhileness at the three major nodes of the production–consumption cycle are eliminated from the list of acceptable solutions.

Further exploration in depth within an engineering framework of economic worthwhileness, with its dependence on time, service life, and obsolescence, is the proper subject matter of engineering economy. The subject will be pursued here only to the extent that an exposition of the design undertaking demands.

Financial Feasibility

Sometimes it happens that a project, meritorious from every point of view, and of great economic worth, cannot be realized because the financial resources for its prosecution cannot be mustered. Public works, depending on bond issues, abandoned when finances are not forthcoming, is a case in point. Similar instances often occur in private enterprise. Before substantial commitments are made for design work the project should be examined for financial feasibility. It may well be that some of the proffered solutions will lead to greater financial demands than others; some may require greater financial resources than are available, and thus be abandoned.

The last three steps are like sieves. Through the first sieve are passed only those solutions which are physically realizable; through the second, only those possessing economic worthwhileness for producer, distributor, and consumer; through the third, only those that are financially feasible. The set of useful solutions comprises the ones passing successfully through each of the three sieves.

In fine, the feasibility study is the first major phase of a design project. It establishes whether the problem posed by the project is soluble and whether there are solutions that are likely to be useful. It is accomplished in a series of design steps any one of which may be repeated, subject to economic considerations, until the interactions of successive steps have been properly accommodated. The first two steps, which emphasize the gathering and organizing of information, are: validating and expanding a primitive statement of need into an engineering specification of outcomes, and formulating the central problem of the project and identifying critical over-all design variables, constraints, relationships, and criteria. The next step emphasizes synthesis and is directed toward evolving a set of possible solutions that are plausible in the light of an understanding of the problem. The next three steps emphasize evaluation and eliminate, in turn, those solutions which are unlikely to be – firstly, physically realizable; secondly, economically worthwhile; and thirdly, financially feasible. The result is a set of useful solutions.

2 M. K. Starr

Modular Production – A New Concept

M. K. Starr, 'Modular production – a new concept', *Harvard Business Review*, vol. 43 (1965), pp. 131–40.

In the course of its normal development, an industrial society begins with an orientation that is focused primarily on production. This is hardly surprising, since technological mastery of a *process* lies at the root of incipient economic progress. Witness, for example, the impact that such process inventions as the printing press, the spinning jenny, the mandrel lathe, and the open-hearth furnace have had on society. During this stage, managers of the production function find themselves actively participating in the topmost managerial affairs.

Then, for evident and fundamental reasons, the situation can change. Production decisions, though recognized as vital, cease to be *critical*. They no longer need to be made at the highest levels of the enterprise. Feasible production alternatives appear much alike, offering little choice. Production decisions become involved with increments of efficiency.

This second stage has more or less characterized the technologies and methodologies of production management for the past twenty-five years in the West. The production function during this period could be sensibly described as 'reliable'. Production decisions have become relatively predictable.

It is a tribute to the pioneers of production management that they have been able to remove great variability – and, consequently, risk and uncertainty – from this area of enterprise. On the other hand, by succeeding in this fashion, they have relegated to other areas, such as finance, research, and marketing, the real *leverage* factors in top management decision making.

But now change is occurring once again. Many enterprise managers are aware that something new is creeping into the production sphere. For a while it seemed as if automation were its name. At least the more journalistic spokesmen, sensing a dramatic change, felt obliged to play on this theme. On careful examination, however, it becomes apparent that automation is only a tool of this new force – and even hardly that in its present form.

The New Capacity

Many production managers are not prepared to understand what is going on. Some hardly even sense that they are in a truly different era. The reasons for this are clear. There is no precedent to go by, and no means for an individual whose interest centers in one functional area to recognize the existence of the trend.

The change that we are talking about can be briefly described as the consumer's demand for *maximum productive variety* (or maximum choice). To achieve this variety, what I call 'modular' or 'combinatorial' productive capacities – that is, capacities to design and manufacture parts which can be combined in numerous ways – are required, as well as compatible managerial abilities.

As has been suggested, the drive toward productive variety is being forced by external factors. It is made possible because of internal ones. An overview is needed for the trend to be perceived. We can state that:

1. The force for this change comes from the market-place.
2. The means for change resides in the production management area, specifically in production's use of: Methodology derived from the management sciences; Technology derived from the physical sciences; Data processing ability obtained from electronic computers.

In other words, the consumer is demanding ever greater variety from which to choose. And new methodologies and technological achievements have developed within the production area which permit the consumer to force this issue (albeit with varying degrees of success).

Irresistible forces

The speed with which such change will occur is influenced by competitive factors. If there is general management resistance, new forms can be expected to grow at the expense of the reluctant. In any case, various alignments of industrial power will arise according to the ability of each organization to assess and modify its own position in an environment that is undergoing change at an accelerating rate.

The problems that must be faced by the next generation of production managers are numerous. Once again production managers are likely to be assigned top-management positions. All industries will not be similarly affected; their adaptations can be expected to follow different timetables. How to speed up or slow down the rate of change will be a subject of interest; but in the long run the forces for change are irresistible, and the degree of change will be recognized as having been predetermined and not under the manager's control.

Accordingly, I shall undertake in this article to describe modular production and to point out where the effects of the developing combinatorial methods are likely to be felt most strongly. I shall try to focus on the kind of managerial training that will be required in order to provide some assurance of success in meeting new challenges as they are raised. The evolution of equipment, alterations in organizational structure, changes in managerial methods – these are some of the other issues I shall examine.

Nature of Metamorphosis

The center of change resides in the production management area, which has long been a stepchild of industrial drama. However, production management cannot swing the change by itself. For those companies that will evolve successfully by maintaining or increasing their total share of a growing market, a new form of effort which will cut across many functional organizational areas is required. We can call this new orientation a 'synthesis' to distinguish it from 'analysis', the euphemism which epitomizes the traditional production management approach.

Production managers are being called on to find (within their *potential* set of tools and concepts) the operational and economic means for introducing real diversity in production output, using a given configuration of plant and equipment. Real diversity means far more than the illusion of an 'adequate' product mix. It signifies the capability to produce a sequence of items (or small batch of items), each of which can differ markedly from preceding and subsequent items (or batches). To achieve differentiation, there will be small batches of units with separate setup charges. And yet, because of the technology and methodology employed, the total charges, including those for setup, can be significantly less than the total revenue obtained in the market-place.

Spotlight on production

It is not difficult to see why production management decisions must be at the core of this activity. Three reasons stand out:

1. The essence of the depicted capability is technological. The notion of high-volume, low-cost, automated mass production will eventually give way to adaptive automation capable of producing a sequence of unique outputs at no sacrifice of volume and at no significant increase in cost. Already, in fact, this trend is in evidence. Thus:

In the automobile industry, a large and still increasing number of ordering options for a new car are readily available to the consumer.

A major petroleum company has designed a gasoline pump that permits the consumer to mix his own blend.

Self-fitting clothing, such as stretch socks, has a kind of built-in diversity.

Increases are occurring in the available variety of soap colors, type faces for typewriters, and shades, styles, and scents for lipsticks, nail-polish, perfumes, sun lotions, and other cosmetics.

There is increased diversity in the size and type of TV sets now available.

In all of these cases, it is growth in technological capability that has permitted such diversity to occur.

2. The new methodology of production management is capable of providing controls for high output diversity. 'Management science', as it is called, has resolved such basic problems as scheduling work, providing adequate inventories, and exercising

quality control, thus permitting splendid managerial control over an enormous variety of products in the product mix.

Significantly, management science has been easily and directly assimilated by only one functional area of business – production management. This observation has been made frequently; yet its relevance has been overlooked. Management science and the production function are natural partners. And as management science unfolds, the production function responds.

3. Because of consumers' attraction to diversity, the ability to produce a real variety offers the kind of high leverage that attracts top management attention. The change represents a fundamental departure from prior conditions in the sense that previously one company's ability to mass-produce identical items could be copied by other firms. But for *modular* production there are no patterns available with which to model and manage the production system. A great deal of innovation and creativity is called for, and with these will be found the inevitable partners, risk and uncertainty, and real vulnerability to astute competitive practices.

Control of variability

Our understanding of the production of real variety can be improved by taking a brief historical excursion. In the early 1900s, students of production management began to learn how to combine many different input elements in a transformation system so as to produce a stream of output units that were sufficiently similar to be interchangeable in every important detail. In effect, to achieve mass production it had been necessary to learn how to remove or control the inherent variabilities of the inputs and the process. This accomplishment can be illustrated as shown in Figure 1.

Outputs from such a production system could be mixed together in a bin and withdrawn in any order for assembly with other parts that had been similarly produced. This gigantic step forward in concept revolutionized the production process. In retrospect, the step seems to have been inevitable. For instance, while Eli Whitney was developing the notion of interchangeable parts in the United States, Leblanc was also making the concept operational in France – but neither man was aware of the other's work. (It may be noted as well that the growth of modular production partakes of the same sense of inevitability.)

The effective realization of the relationship shown in Figure 1 was the result in no small measure of the efforts of such men as Frederick W. Taylor, Henry L. Gantt, and Frank Gilbreth. Few managers, no matter what their functional area, are unaware of

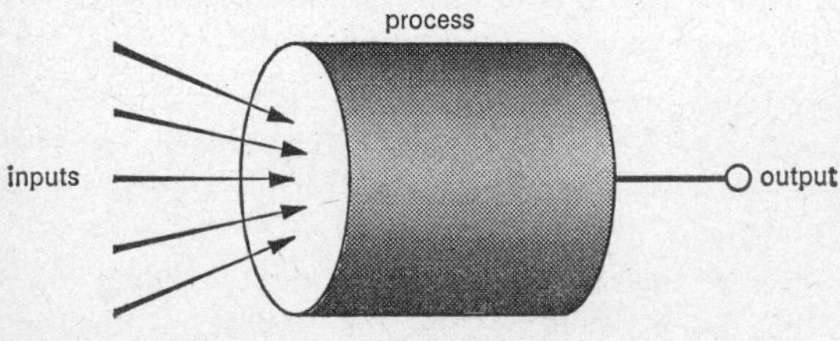

mass production requires control (and reduction) of the variabilities that are inherent in any set of inputs (e.g., materials, workmanship) and in the manufacturing process. In this way, a stream of identical units of output can be obtained

Figure 1 Control with mass production

the objectives and achievements of these early pioneers. Somewhat less well known, but totally appropriate to this discussion, is the fact that when it was realized that some variability in the outputs must always exist, Walter Shewhart and his colleagues developed the function of statistical quality control, whereby the level of fundamental or inherent variability could be described and brought under management direction.

As a final step, Henry Ford helped to complete production's conquest over variability by providing an operational instance of co-ordinated assembly for mass production. The result was that the multiple inputs could be transformed in both space *and time* to afford a controlled stream of output.

With this background in mind, let us turn from developments in the factory to trends in the market-place.

Demand for Variety

We know that the demand for variety originates at the consumer level. But how much variety does the consumer want? How much is he willing to pay for it?

In the past, operating under a production configuration of the type illustrated in Figure 1, marketing management stepped into the breach and accepted the challenge to supply the consumer with apparent variety. Production facilities could not generate as much *real* variety as the market was able to absorb, for the technological capability was lacking, and so was the managerial ability to control the production of diverse outputs. So the marketing function, aimed at diversity, began where mass production (indicated by the single output line in Figure 2) left off, and its success is amply illustrated by the important status of marketing in management in recent years.

In general, marketing activities directed toward creating variety have represented an increasing proportion of the total costs of an item. The consumer has not been reluctant to accept these services.

Increasing expectations

Initially, the marketing operation was most effective when the diversity of production was minimal. Each firm established some unique qualities for its brands. The consumer was able to contrast these attributes with those of other brands. Differentiation was based on many subtleties, not the least of which were the name, the package, and the image associations that were created. One basic chemical mixture could be positioned in the market-place in several totally different ways. Even a single brand might appear to possess quite diverse qualities to different demographic segments of the population if the advertising and promotion could be properly controlled.

But such differentiation was not to be enough. Pressure exerted by the consumer for greater variety began to appear in several different forms. For instance, the consumer began to question the 'truth' of apparent variety. As a result, marketing, advertising, and promotion operations have come under the critical surveillance of consumer groups and the government. It appears, however, that disillusionment about brand marketing, to the extent that it exists, has *not* been connected with the purpose of marketing or with the concept of apparent variety, but with the lack of a commensurate measure of *real* variety. The consumer is simply asking for a more significant choice.

Pressure for variety also helps to explain the severe contraction in product life that has been experienced in recent years, and the unusually high failure rate for many new products which require acceptance by a large market segment for economic success, yet are not sufficiently unique to warrant such acceptance. It is also

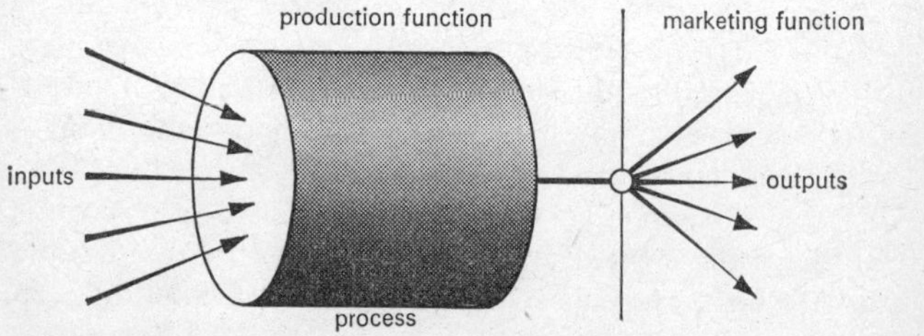

Figure 2 Product variety with mass production

significant that the demand for variety appears to follow a cyclical pattern. The oscillations are correlated with production capabilities and the various phases of technological maturity. The consumer's expectations appear to be entirely in keeping with the state of the art:

As technology advances, the first reaction of the market-place is to adopt the prototype model of the new technology as widely as possible. Only minimum brand differentiation and almost no variety within a single brand are found. The issue of primary importance to the consumer is the *ownership* of the car, radio, or other product.

Then, as soon as the prototype models are broadly distributed among consumers, a shift begins to take place. Various market segments appear, and the accumulation process proceeds in terms of particular brand names and special model numbers.

Finally, the market-place begins to discard the notion of 'keeping up with the Jones' and emphasizes instead the uniqueness of the individual's possessions (Dichter, 1965). This trend is based on the fact that the quality of difference has value for the consumer. Acceptable difference is not, of course, a random affair but is based on some personal predisposition. 'Keeping up with the Joneses', if it has any meaning, must be interpreted as being acceptably different from the Jones where the difference can be explained in terms of a purely personal and individual philosophy.

Production meets the need

Marketing managers have succeeded to *some* extent in presenting the consumer with a choice so that he can exercise his individuality. In part, this has been achieved by stressing non-functional factors of quality which can serve to distinguish one product from another. But the consumer's drive for uniqueness appears to be insatiable. For instance, the novelty of European cars led to an unexpected degree of acceptance by U.S. consumers. The capricious affair of the U.S. consumer with compact cars has been of the same type. And the great increase in consumer spending for such services as travel and entertainment can also be accounted for as another way to further one's individuality.

What makes all this so significant is the timing. The drive toward highly personalized possessions comes at a time when the production manager begins to find himself in a position *to deliver this real diversity* without violating reasonable economic bounds. The degree of personalization that is required is more than that delivered by monogrammed luggage, shirts, ties, and automobiles. It is more than what 'key clubs', unique shapes and colors of telephones, or antique household furnishing can provide. It is embodied in the growing success of mail-order variety, in the concept of 'do it yourself' (with a great many options to do it 'in your own way'), and in other forms which we shall mention presently.

Beginnings of the Trend

The kind of timeliness that we observe here is not unusual. There is, if anything, distinct precedent for it – production has responded to market demands before. Only the specifics are different.

Elements of change

To appreciate the new trend, one can divide the production system into two basic parts:

1. The transformation processes.
2. Assembly operations.

Many different inputs (including materials, skills, information, and power) enter the transformation process. They are combined in various ways to produce a catalog of parts. *It is the essence of the modular concept to design, develop, and produce those parts which can be combined in the maximum number of ways.* Figure 3 is intended to present visually some of the aspects of this statement.

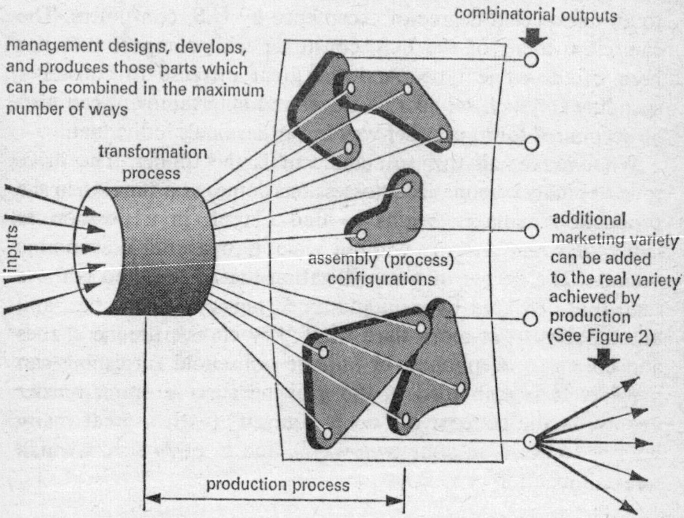

Figure 3 Concept of modular production

This emphasis on maximizing the combinatorial variety of assemblies from a given number of parts is the new element of interest to production managers. It is what is meant by modular capabilities. One accent is on design; a second is on the exercise of adequate managerial controls.

There are numerous examples that can be given of systems which have developed with a *tendency* in this direction – jet engines, originally designed for airplanes, being adapted for use as integral components of power plants; modular office wall panels and furniture permitting a wide diversity of arrangements, and the similar configuration adaptability of prefabricated structures. But none of these examples can suffice to describe what would result from the total planning of a system of this kind; the production manager can now deliver a much greater variety, at a reasonable cost, than his present activities indicate.

Retarding factors

At present, there are a number of difficulties that stand in the way of the development of advanced modular production:

1. Many production managers are unaware of the market potential for highly diversified outputs. Accordingly, they have not familiarized themselves with the technological problems of combinatorial design which are involved.

The recent acquisition of new methodological abilities, however, is forcing production men to examine implications for managerial control of richly diversified systems. And the enormous manipulative abilities of electronic computers may reinforce the feeling that some new potential has entered the picture. In particular, recent developments in designing products by computer and the use of visual input–output peripheral equipment aids, such as those employed by the General Motors Corporation, have great significance for the achievement of sophisticated combinatorial variety in the design of modular parts. These computer systems can store design components such as rectangles, circles, force fields, and stress patterns. Indeed, a great variety of component combinations can be tested by means of analytic and simulation procedures, eliminating the need to construct a prototype model for physical testing.

2. Many marketing managers, though aware of the market potential for highly diversified outputs, are convinced that their traditional approach to the consumer is sound. Without a complete revision in perspective (which inertia prevents), the new potential goals cannot be defined. And one cannot measure

how far short he falls from an unspecified goal. As a result, the demands of the market-place have not been translated into production terms.

(Judging from all indications, research and development managers often suffer from the same difficulty.)

3. From the viewpoint of society, modular production will bring with it significant costs in the form of obsolescence. Obsolescence occurs not only with respect to the present product line but also with respect to present configurations of plant and facilities. While it can be argued successfully that changeover to modular production need not be total or swift, the fact remains that turning to the modular approach produces a great deal of unplanned obsolescence. In addition, the design and engineering costs of entering into such production configurations can be exceptionally high.

Management personnel must be trained to cope with both the concepts (for planning) and the realities (for operating). The background essential for this training will be the management sciences. In particular, the combinatorial aspects of mathematical analysis will provide an essential foundation – above and beyond which managerial intuitions can and will continue to be exercised.

In view of the uncertainties, new risks, and difficulties, it is predictable that many otherwise alert executives will tend to look the other way when an issue pertaining to real productive variety is raised. To wait and see, to postpone action until one's hand is forced – these attitudes will tend to be substituted for what is ordinarily called entrepreneurial wisdom.

Pressures for change

Meanwhile, however, there will be situations where reluctance to move in the new direction will seem absurd to businessmen involved. The totally new venture, the new or experimental autonomous branch, and industries that traditionally are involved with substantial retooling – these are exemplary situations for the introduction of modular capabilities. In time, such instances may provide the momentum to force still other hands. It seems logical to suggest, as well, that high-volume industries characterized by short product life cycles will react early to the possibilities of modular production, and that industries which characteristically are involved with expensive parts and components will not overlook the potential benefits of modularity.

Only in some ultimate sense can we expect that the *over-all* industrial system of the present will evolve to the point where it becomes truly responsive to the selection process of the individual. What we *can* anticipate is the continued fragmentation of the market to ever smaller units that can be serviced with relatively more unique outputs, with *some* enterprising companies capitalizing on the trend.

Planning and Control

At this point I shall depart from discussion of general trends and problems in order to demonstrate in a more precise way what modular production means to management. This demonstration will allow us to draw some important conclusions for those men entrusted with planning and control.

Parts and configurations

A better understanding of the nature of modular capabilities for real productive variety can be obtained by constructing a table of the kind shown in Table 1. We see that, with N different kinds of parts, a total of M different product configurations is derived. Some of these products require several units of a single part; for instance, PR_1 requires two of PA_2, and PR_j requires two of PA_i. The maximum possible variety, i.e. maximum M, that can be obtained with N different parts may be very large. This is especially so when we include both the different possible combinations of the parts and varying numbers of them in combination. Thus:

If each part can appear zero, one, or two times (i.e. three options), then the maximum variety that can be derived with N parts will be $3^N - 1$ (one is substracted because there will be one configuration in which none of the parts appears). Considering only five parts ($N = 5$), the maximum variety (defined in this way) will be 242.

Table 1

Specifications for the Kind and Number of Parts to be Combined for Each of a Given Number of Different End Products

Variety of parts	*Variety of products*					
	PR_1	PR_2	PR_3	PR_4	PR_j	PR_M
PA_1	1	0	1	1	0	0
PA_2	0	1	1	2	0	0
PA_3	0	0	0	0	1	0
PA_4	0	0	1	0	0	0
PA_i	0	0	1	0	2	0
PA_N	0	0	0	1	1	1

Notes:

PA_i denotes the part identified by the stock number *i*.
PR_j denotes the product variation listed in the finished goods catalog as *j*. (As shown above, the product *j* assembly requires one unit of part 3, two units of part *i*, and one unit of part *N*. The sequence of assembly is not indicated.)
PA_N denotes the last part listed in our table.
PR_M denotes the last product variation listed in our table; in this case, it is the product using only one unit of part *N*.

With ten parts, the maximum variety rises to a somewhat astonishing 59,048.

If, in addition, the ways and sequences with which the parts can be combined also add variety, then the number of possibilities rises even faster.

Of course, many of the theoretical possibilities cannot exist and would have no appeal as a product choice for any consumer. But the basic idea of modular design is to have an inventory of parts which can partake in many appealing product configurations. Each order received by the company is translated into a unique assembly configuration. (The same sort of table as shown in Table 1 must have been designed long ago for the first Erector set. It appears that industry is about to take seriously a fundamental principle long embodied in many children's toys.) A useful measure of the effective degree of modular design might be, therefore, the number of products (columns) that can be generated from a given number of parts (rows).

Economic model

It is now possible to expand the notions that we have been developing into an economic framework. As a convenient approximation, let us begin with the assumption that we can produce fixed but different amounts of each part and that we have *N* such parts which can be assembled in various ways to yield a variety of *M* products. If we know the costs of each part, we can combine all of this information by means of a linear programming model to determine an optimal product mix. The objective would be to maximize profits, subject to the constraints imposed by the modular system of parts that has been pre-designed and by the productive capacity of the enterprise to produce these parts.

From the fundamental theorem of linear programming we know that the number of parts (as we have defined them) plus the number of any other resource constraints (such as storage space, limited power, or scarce raw materials) will be equal to the maximum number of products that can be included in the ultimate composition of the product mix. But this rule produces the somewhat absurd result that the product mix will be only slightly larger in size than the total number of parts that have been designed by the company. Such a result would defeat the purpose and concept of modular productive capability.

What, then, can account for the fact that many more product configurations than are indicated by the far smaller number of parts would, in fact, be included in the ultimate composition of the product mix? The answer must lie in the demand of the market-place for uniqueness. Once we have accepted the fact that the revenue curves for each product configuration are not linear but fall off as the uniqueness of the product class diminishes, we have a satisfactory explanation for expanding the product mix.

Figure 4 compares a linear revenue curve with one that decreases as the uniqueness of the product class in the market-place diminishes. The linear form represents complete consumer indifference to the amount and scope of product distribution.

The nonlinear curve of diminishing returns, on the other hand, indicates decreasing consumer responsiveness as the extent of distribution and ownership of a particular product configuration increases. (The decline in responsiveness may be hastened by repeated purchases over time.)

Optimal product mix

The saturation phenomenon is well known in marketing. It is usually based on the notion of an exhaustion of 'prime' targets. I am suggesting that the size of the potential market diminishes at a faster rate than can be accounted for by exhaustion alone, or, in other words, that the acquisition of a new customer will remove more than one potential customer from the market-place. The reason is that every customer added increases the consumer's disenchantment with what he conceives to be boring, commonplace, or ordinary.

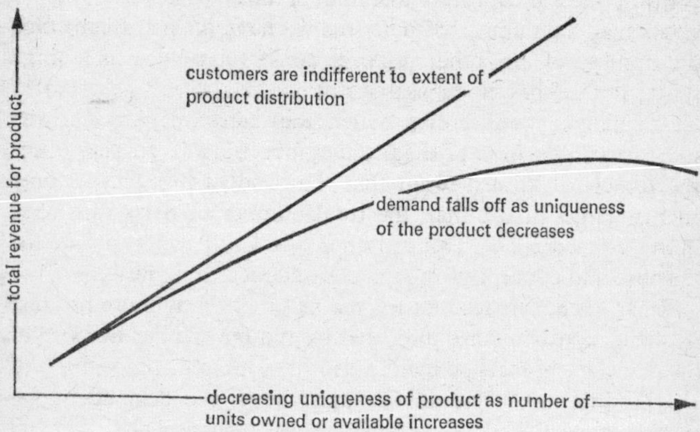

Figure 4 Linear and nonlinear revenue curves

When a nonlinear form of the kind shown in Figure 4 is encountered, the effect with respect to a linear program is to increase the number of constraints. As a result, the size of the optimal product mix increases. Under such conditions, and even with relatively few parts, a great number of product configurations will be required in order to maximize revenue. The more demanding the consumer is for uniqueness in his possessions, the greater the curvature of the revenue lines. And in turn, as the curvature of a revenue line increases, the indicated size of the optimal product mix grows larger.

Interchangeable modules

In terms of Table 1, the traditional concept of interchangeable parts can be interpreted as follows: All units that belong to any particular parts classification can be treated as being identical. It does not matter whether a given unit of a certain family of parts was made at the beginning or end of a production run. Any member of this family of parts can combine with equal ease with any member of another family of parts in the row. Note that this interchangeability exists *within* the rows of the table and not between the rows of the table.

Interchangeable part *modules*, on the other hand, are designed so as to be highly transferable *between columns*; i.e. an interchangeable module enters with equal ease into many different product configurations. In contrast, production as it is conceived today tends to isolate the columns.

Once the concept of interchangeable modules becomes widely accepted, it may well be responsible for developments that are astounding or at least as significant as those which followed from the general adoption of the concept of interchangeable parts.

Meeting the Challenge

The need and the ability to provide real diversity will certainly vary in many ways. However, management can anticipate some of the important factors that are likely to operate to the advantage or disadvantage of a company, and plan accordingly.

Changing opportunities

Some of the more likely factors of differentiation will be discernible in terms of the characteristics of industrial sectors, demographic segmentation, competitive structures, organizational flexibility, executive attitudes, and a variety of exogenous factors which are imposed by the economy and international trade. Since starting ventures will possess advantages over going interests with established production systems, the latter may find themselves being pushed at unexpected times and in unanticipated ways.

Organizations seeking to diversify will encounter new opportunities to branch out. Creating real productive variety – that which comes with modular production – is equivalent to building into the system preplanned potentials for diversification. In a related sense, activities such as long-range planning and master planning necessarily will span new horizons and assume new dimensions.

The market-place will cease to be treated as a statistical phenomenon that can be described by gas-law models, analogs of random walks, quantum mechanics, or other such models used for deriving average responses and occasionally for obtaining measures of variability. Instead, there will be a growth of interest in the behavior of the *individual.* The market-place will be most appropriately described as the sum of individual behaviors.

The concept of the highly segmented market will be found to apply quite well to common market-type operations, where consumer diversity occurs because of the structural variants in the environment (e.g. broad cultural differences between consumer segments, and significant production variations caused by price and wage differentials, previously established underwriter codes, and separate systems of national taxation).

Innovation in production

The essential ingredients of change for the production manager are twofold. On the one hand, a veritable revolution in design ability is required in order to move in the direction of interchangeable modules. The roots of such efforts already exist in the well-developed concept of standard parts (e.g. screw threads, gauges, and light bulbs) and in standard process operations (e.g. synthetic time standards and computer-controlled equipment). On the other hand, it will be recognized that adaptive automation rather than a fixed-output type is the ideal to be desired. However, the primary significance of adaptive automation will be for assembly operations and only secondarily for the transformation processes.

The use of the computer seems to be the key. It will be a major asset when it comes to coping with the production of real variety. From another point of view, it is also a force for change. When employed in conjunction with appropriate models from the management sciences, such as sequencing models, programming methods, line-balancing and queuing models, and heuristic procedures of many kinds, a new level of managerial control of diversity can be achieved.

Radically different design techniques can also be expected. Modular design is a realistic objective in the light of visual-aid inputs to computers. These peripheral equipments permit almost instantaneous assessment of a designer's work. Complex design constructions and evaluations that formerly required many man-weeks of analysis will be achieved so rapidly that a new sense of creativity will develop.

Efficiency in distribution

The distribution problem which arises as a result of increased variety is certain to be enormous. Here, too, the role of the computer is likely to be crucial. The rapid communication of individual consumer demands to the production control units, the translation of these demands into specifications appropriate for the transformation and assembly processes, and the distribution of production outputs to the appropriate consumers – all of these steps appear to be feasible because of new insights and abilities *and* the computer.

Organizational steps

It is apparent that we are dealing with a situation that will involve fundamental changes in the enterprise. What is required is the ability to manage a new kind of productivity. An appropriate organizational structure would provide greater responsiveness to the market. In such an organization, production would be in touch with consumers, with the contact mediated by the marketing function. An appropriate organization would also allow the production manager to respond with sensitive perception to developing technologies. To achieve such results, a much higher level of functional integration is called for. It is also quite clear that production management should once again participate in top management decision making.

Product and Project Development

Simulation methods can be expected to play a crucial role in the development of an appropriate organization. They will provide necessary information, as well as assurance concerning the effect of proposed changes in relationships and procedures. Also, when management simulates the design of modular parts, their assembly into many unique products, the managerial control problems that result, and the distribution problems created by a highly segmented market, it can examine the important characteristics of the total system. Thus, simulation provides a rational means for analyzing an enormously complex system whose diversity promises not less but more satisfaction born of less conformity.

References

DICHTER, E. (1965), 'Discovering the "Inner Jones"', *Harvard bus. Rev.*, vol. 43, no. 3, p. 6.

3 W. J. J. Gordon

The Operational Mechanisms of Synectics

Excerpt from W. J. J. Gordon, *Synectics: The Development of Creative Capacity*, Harper & Row, 1961, pp. 33–6.

Synectics defines creative process as the mental activity in problem-stating, problem-solving situations where artistic or technical inventions are the result. I use the expression 'problem-stating, problem-solving' rather than merely 'problem-solving' in order to include the definition and understanding of the problem. The operational mechanisms of Synectics are the concrete psychological factors which support and press forward creative process. The mechanisms do not pertain to the motivations for creative activity, nor are they intended to be used to judge the ultimate product of an aesthetic or technical invention. Psychological states such as empathy, involvement, play, detachment, and use of irrelevance are basic to the creative process but they are not operational. The Synectics mechanisms are intended to induce appropriate psychological states and thus promote creative activity.

Words like intuition, empathy, and play are merely names put to complex activities in the hope that the naming of the activity will in fact describe it. Experience has shown it to be most difficult to feed back into a problem-stating, problem-solving situation such nominalistic abstractions. When dealing with an individual or a group faced with problem-stating and problem-solving, it is ineffectual to attempt to persuade the individual to be intuitive, to empathize, to become involved, to be detached, to play, or to tolerate apparent irrelevance. However, in our research experience the Synectics mechanisms effectively increase the probability of success when creativity is called for. They draw the individual into the psychological states.

The Synectic process involves: (a) making the strange familiar; (b) making the familiar strange.

Making the strange familiar. In any problem-stating, problem-solving situation, the first responsibility of individuals involved is to understand the problem. This is essentially an analytical phase where the ramifications and the fundamentals of the problem must be plumbed. However, if only this analytical step is taken, no novel solution is possible. For work on a problem to get started, some concrete assumptions must be made, although in the course of the problem-stating, problem-solving process, the understanding of the problem may change. It is the function of the mind, when presented with a problem, to attempt to make the strange familiar by means of analysis. The human organism is basically conservative, and any strange thing or concept is threatening to it. When faced with strangeness the mind attempts to engorge this strangeness by forcing it into an acceptable pattern or changing its (the mind's) private geometry of bias to make room for the strangeness. The mind compares the given strangeness with data previously known and in terms of these data converts the strangeness into familiarity.

This is, of course, an obvious part of problem-solving. However, Synectics is an attempt to describe those conscious, preconscious, and subconscious psychological states which are present in any creative act. Therefore, it would be an omission not to mention the analytical, the making-the-strange-familiar mechanism. The great pitfall, the traditional danger, in making the strange familiar is in becoming so buried in analysis and detail that these become ends in themselves, leading nowhere. The process of making the strange familiar, if used alone, yields a variety of superficial solutions; but basic novelty demands a fresh viewpoint, a new way of looking at the problem. Most problems are not new. The challenge is to view the problem in a new way. This new viewpoint in turn embodies the potential for a new basic solution.

Making the familiar strange. To make the familiar strange is to distort, invert, or transpose the everyday ways of looking and responding which render the world a secure and familiar place. This pursuit of strangeness is not a blasé's search for the bizarre and out-of-the-way. It is the conscious attempt to achieve a new look at the same old world, people, ideas, feelings, and things.

In the 'familiar world' objects are always right-side-up; the child who bends and peers at the world from between his legs is experimenting with the familiar made strange. (One sees the familiar tree as a collection of solids in an otherwise empty space. The sculptor consciously may invert his world and see the tree as a series of voids or holes carved within the solid block of the air.)

Owen Barfield quotes a South Sea Islander's pigeon-English description of a three-masted, screw steamer with two funnels: 'Thlee-pieces bamboo, two-pieces puff-puff, walk-along inside, no-can-see'. In our terms, the conceptions which frame the steamship are firmly established in the realm of the familiar. Here, the familiar Western concept of steamship is juxtaposed with the strange pigeon-English version. Barfield says,

> Now when I read the words, 'Thlee-pieces bamboo, two-pieces puff-puff, walk-along inside, no-can-see', I am for a moment transported into a totally different kind of consciousness. I see the steamer, not from my own eyes, but through the eyes of a primitive South Sea Islander. His experience, his *meaning* is quite different from mine, for it is the product of different concepts. This he reveals by his choice of words; and the result is that, for a moment, I shed Western civilization like an old garment and behold my steamer in a new and strange light (1957, p. 49).

The steamer seen by the Western mind in this light is reconstituted and presented as alive and malleable to the imagination.

These several mechanisms for making the familiar strange are not a collection of mental tricks for the achievement of superficial novelty. They have been developed and are used in the several Synectics groups as a systematic way of solving actual invention problems. Thus, Barfield's 'new look' at the familiar steamer could be the starting point for considering propulsion. Combined with technical competence it could in turn lead to a new viewpoint for the development of a new invention.

The attempt to make the familiar strange involves several different methods of achieving an intentionally naïve or apparently 'out of focus' look at some aspect of the known world. And this look can transpose both our usual ways of perceiving and our usual expectations about how we or the world will behave. The experience of sustaining this condition can provoke anxiety and insecurity. But maintaining the familiar as strange is fundamental to disciplined creativity. All problems present themselves to the mind as threats of failure. For someone striving to win in terms of a successful solution, this threat evokes a mass response in which the most immediate superficial solution is clutched frantically as a balm to anxiety. This is consistent with the natural impulse to master the strange by making it familiar. Yet if we are to perceive all the implications and possibilities of the new we must risk at least temporary ambiguity and disorder. Human beings are heir to a legacy of frozen words and ways of perceiving which wrap their world in comfortable familiarity. This protective legacy must be disowned. A new viewpoint depends on the capacity to risk and to understand the mechanisms by which the mind can make tolerable the temporary ambiguity implicit in risking.

References

BARFIELD, O. (1957), *Poetic Diction*, Faber.

4 D. A. Schon

Uncertainty is Inherent in Technical Innovation *and* The Cost of Innovation

Excerpt from D. A. Schon, *Technology and Change: The New Heraclitus*, Delacorte Press, 1967, pp. 24–37.

Uncertainty is Inherent in Technical Innovation

Men involved in technical innovation in a corporation confront a situation in which the need for action is clear, but there is both danger and opportunity, and it is by no means clear what to do. This situation is painful and anxious for individuals and, in a sense, for the corporation as a whole. To be a member of a corporation in a time of corporate uncertainty is to be engulfed in waves of uncertainty and anxiety. As long as this situation obtains, the corporation cannot function effectively. The corporation is not designed for uncertainty, where there are no clear objectives to reach and no measures of accomplishment, and where it is not clear what to try to control. A corporation cannot operate in uncertainty, but it is beautifully equipped to handle risk. It is, from at least one point of view, precisely an organization designed to uncover, analyze, evaluate, and operate on risks. Accordingly, *the innovative work of a corporation consists in converting uncertainty to risk*. This work begins with more information than can be handled and operates on this information, at lower levels of the corporation, until clear alternatives of action, together with their probable benefits and their risks, can be defined. At this point, management can play the investment game, that is, the game of deciding where to put one's bets. The game requires analysis of investment alternatives, estimating markets, costs and technical feasibility, and making investment decisions. The game is played with competitive corporations as opponents. The rewards and punishments of the game can be measured in dollars.

In the process of innovation everything is done to permit decision on the basis of probable dollar costs and dollar benefits.

In the process, the corporation converts the language of invention to the language of investment. Instead of talking about materials, properties, performances, experiences, experiments, and phenomena, the corporation begins to talk of costs, shares of market, investment, cash flow, and dollar return. The language change accompanies a shift upward in the level of corporate involvement.

The conversion of uncertainty to risk takes varying forms depending on the *kind* of uncertainty to be dealt with. Technical innovation involves many different kinds of uncertainty. Some of these spring directly from the nonrational character of the process of invention. Others are only indirectly related to that process. We will be concerned here with uncertainties springing from determination of technical feasibility, novelty and market.

One focus of uncertainty is in the question, Can it be done? Is it technically feasible?

In the process of invention, as we have seen, the requirements to be met change continually in response to unexpected findings. These may at any time turn a good risk into a poor one or an indifferent idea into an idea of great promise. We can describe these changes in terms of a curve of technical difficulty.

As an investigator works he may first encounter the most difficult problem and later only minor ones whose solution he knows he can attain if he works long enough. If we plot his progress in time against 'difficulty' or 'demands of the solution with respect to the state of the art', the picture looks like this:

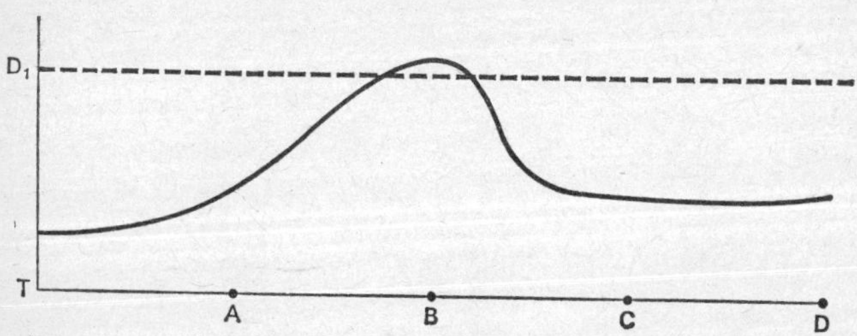

Figure 1

The dotted line at D_1 represents the state of the art. The investigator has to surpass it at B (for example, by developing a wire coating with much improved holding strength), whereas his subsequent tasks are well within the state of the art. On the other hand, there may be a number of separate peaks. In considering the development of new semiconductor materials, we would have to identify separate peaks for the development of high-purity materials and for the development of high-yield production techniques.

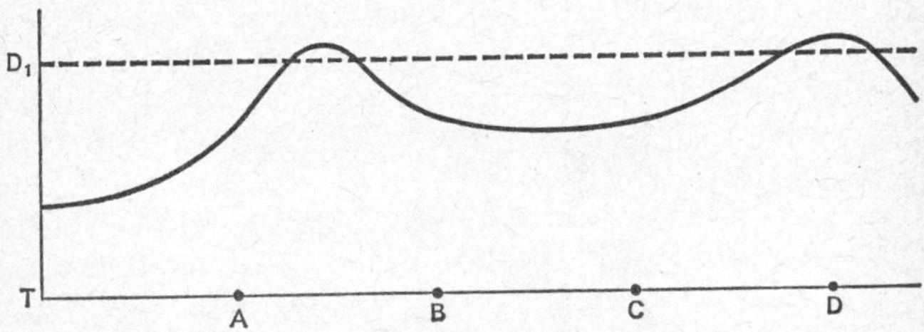

Figure 2

Such curves would not be a source of major uncertainty if it were possible always to identify ahead of time the problems and their degree of difficulty. For example, small improvements in the properties of materials – raising by a few degrees the melting point of an alloy or increasing slightly the abrasion resistance of a plastic coating – may require no more than the continued application of known methods. On the other hand, achieving these small improvements may be like approaching absolute zero. The investigator may be fighting the equivalent of the basic laws of thermodynamics. For example, a group working on the development of a new type of transformer, using piezoelectric principles, worked for weeks on what it thought was a materials problem, until it was able to make the calculations necessary to show that the performance requirements for which it was striving pressed to the limits both the energy capacity of the material and its ability to accept mechanical stress.

It is not always apparent, even to a skilled investigator, whether he is working on a minor problem of adjustment or a major problem of principle. He cannot place himself on the curve of technical difficulty. He may think he is here:

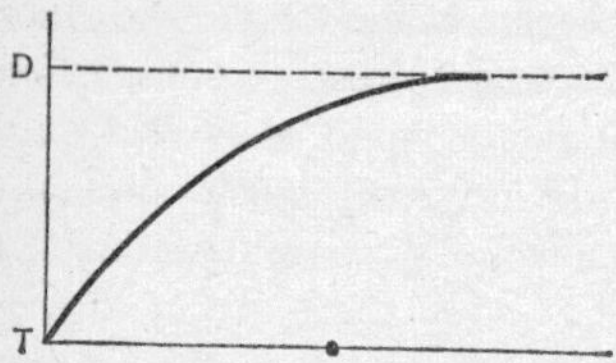

Figure 3

when, in fact, he is here:

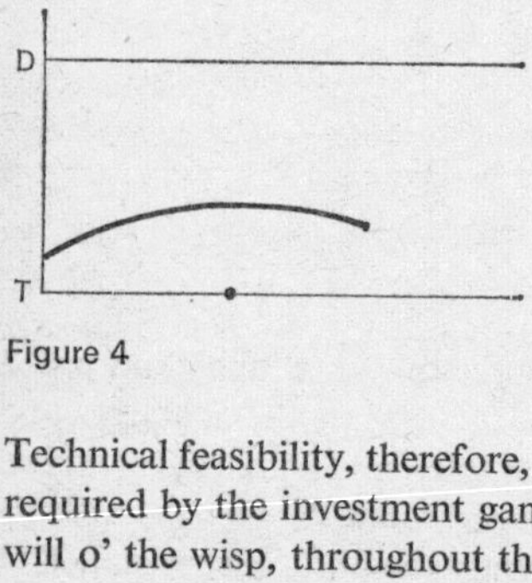

Figure 4

Technical feasibility, therefore, often resists the kind of definition required by the investment game. It may evade definition, like a will o' the wisp, throughout the entire process of innovation.

Another focus of uncertainty concerns the question of novelty: Who has done it before? or, Who is doing it now?

Although the patent literature offers a great deal of information about prior invention, a search may not uncover all relevant art. Even a thorough and concentrated search may not turn up a relevant configuration which had been invented earlier for quite a different function. An apparently novel screw thread may have been used in a machine tool to fasten work to a chuck. A method for encapsulating adhesive may have been used for encapsulating solid rocket propellant. Such facts may come to light, in spite of the best efforts of the searchers, only when the patent application is processed or later when a counterclaim is filed. If someone contests a patent, the courts often find against the alleged inventor. The attitude of the courts toward patents, the skill of the attorneys and the credibility of the witnesses may count for as much as the apparent novelty of the invention. The novelty of an invention is uncertain, both because of lack of information, in spite of best efforts, and because of the uncertainties of the judicial process.

If a company undertakes a development, it can have no guarantee that its competitors are not doing similar things. Even if it knows that others are working in the same field, it cannot be sure how far they have gone. Invention is like an arms race between warring countries: The leaders ask: Do they already have it? Are they working on it? How soon will they get it? Can we afford not to have it? Should we wait until they have it and then copy it? What do we lose by being second? How much would we gain by being first?

In corporate competition the same questions are relevant. There are the same difficulties in getting answers, the same temptations to espionage, the same strategies for playing out a hand in spite of uncertainties. A company may be well advised to continue its development, even with the knowledge that a competitor is working on the same problem: it may get there first and the other may not get there at all. There are even cases on record – the development of the automatic light meter for cameras is one – where a company, spurred on by the belief that a competitor had the product, finished first and achieved a commercial success – and discovered that its competitor did not have the product after all.

Who will buy the product? How large will its market be? How much of a share will it get? How long will the market last?

The marketing questions have become the subject matter of a new profession and of what aspires to be a science. Without attacking the question of whether or not a marketing science is now possible – whether there are inherent uncertainties about marketing which no amount of information, ahead of time, will resolve – we can notice that the process of answering marketing questions occurs in time and that, as in the process of invention, there are bound to be stretches of uncertainty which can only be resolved by the expenditure of time and money – if then.

Scotch tape was originally introduced for mending books. Before its actual introduction, its makers had no inkling of the huge market potential it offered.

Vinyl resins were first developed and marketed as adhesives. Their use as structural plastics was only dimly foreseen.

These success stories suggest that products are vehicles for projection: what they are depends on how you see them. It is possible to see in them more or less than their makers intended. In the examples above, buyers saw in the products more than the makers had originally perceived. There is no lack of examples in which they saw less.

The uncertainties of marketing are particularly apparent in the case of consumer products. A whole industry – the garment industry – rests on the concept of fashion and yearly fashion change. The process of taking orders, establishing inventories and making production schedules depends on anticipation of what the trade will want – within the lead times required for the production process. And even the most skilled practitioners regard the anticipation of fashion trends as an art, an intuitive skill, which can best be done by the seat of the pants. In many industries market trends remain obscure until the yearly trade show, and then, of course, there is no guarantee that wholesale buyers' judgment will be reflected in consumer behavior. The number of industries subject to fashions of their own goes far beyond the few – garments, cosmetics, jewelry, footwear, etc. – which are clearly seen as a fashion business.

Marketing is, more than any other, the process in which there is more information than you can handle. Although you may be able to predict, in principle, the effects of single-variable changes – price, for example, or color – these variables never act alone. There are always many relevant variables, with multiple, interacting effects.

A company tried to market in California a food product which had sold well in Detroit. The result was disastrous. Why? Detroit was familiar with opaque packaging, California with clear film. In Detroit the idea of dietary snack foods was apparently acceptable; in California, it was not. Or had the product been acceptable in Detroit in spite of its dietary connotations?

In Detroit the product had enjoyed decades of use; its trade-mark was well known. In California, it was relatively unknown, but the company spent hundreds of thousands of dollars promoting it. Billboards and television 'spots' were used in California. Was the failure due to the inappropriateness of these media? Were there quality problems in the California plant?

In theory the company could identify and test each of the variables possibly responsible for failure. In this way they might unravel and solve the problem. But there was not sufficient time. Each day, tens of thousands of dollars were slipping away. Something had to be done quickly *without a clear definition of the problem.*

Companies make marketing decisions under pressures of money and time, confronted with more information than they can handle. There is no sure way of learning from past experience. The situations are never identical – California is never quite Detroit – and apparently trivial differences may turn out to be critical.

Moreover, it costs money to find out. The introduction of a consumer product on a nationwide scale is a major enterprise. A single regional market test, with appropriate preparation and follow-up, may cost hundreds of thousands of dollars, and its results may not permit formulation of the national picture. Although uncertainties may be resolvable in principle, the cost of their resolution is high. The very process of resolving them may cost more than can be justified.

To this we must add the fact, outlined above, that in the process of development, need and technology do interact. The market originally conceived for the product may be ruled out by technical limitations discovered in the process of invention. A more limited, a broader or a different market may suggest itself. In short, the product for which a market must be anticipated is not an 'it' that remains constant throughout the process of development. The product changes and its possible market changes with it.[1]

1. The market is in this sense a function of the product, but the nature of the function is never clear ahead of time and may remain unclear throughout the product's life. The properties of the product determine its market but

The Cost of Innovation

Throughout the process of technical innovation decisions about technical feasibility, novelty, and markets, among many other factors, must be made on insufficient evidence, by individuals faced with more information than they can handle. The resolution of these uncertainties – the conversion of uncertainty to risk, which is the corporate work of innovation – takes time and money and requires justification in its own right.[2] Its benefits must be balanced against its costs. The process of innovation has a cost curve, as well as a curve of difficulty, which exhibits characteristic patterns.

Let us consider the development of a new metal-to-metal adhesive. It begins in the laboratory, when a chemist, working on another problem, notices an adhesive effect and reproduces it. The cost, in man hours and materials, up to this point may be in the order of $5000 to $10,000.

The chemist shows the adhesive to the research director, who finds it intriguing and authorizes a search of the report and patent literature, some further experimentation and a first examination of markets. This may go on for a month or two and bring the total cost to $40,000.

At this point, the research director feels he has something

---

in obscure ways. An awareness of something like this underlies the unwillingness of companies to alter a successful product; it has worked, but no one knows why.

The market situation is one in which corporations would like to experiment. But there are few meaningful experiments that can be performed without changing the market situation itself, sometimes in irretrievable ways. For example, an experiment on a small population may be discarded because it is not representative, but experiment with a large population costs a great deal – and may significantly affect the product's future. There are, then, inherent limitations on the company's ability to discover 'the truth' about consumer response to product and to change in product.

2. The uncertainties of market are inherent not only in the structure of consumers' attitudes, which change in complex ways over time, but in the dynamic situation of the firm confronted with the need to find out. Characteristically, the problem is not simply to discover the truth but to provide a basis for decision within definite limits of time and cost. The question is never simply 'What can we find out?' but 'What can we find out within these limits of inquiry?'

worth presenting to management. He makes the presentation and impresses management, which launches a full-scale development project. The adhesive formulation has certain disadvantages, of course. The development team makes efforts to improve it, as well as to explore variations of the chemistry involved in order to cover 'all the ground'. A detailed analysis of the market – in the automobile, aircraft, and appliance industries – gets under way. It reveals the need for improved handling properties. At the same time, the company undertakes a thorough patent search which suggests that, although the formulation may be novel, the company must try certain formulations as well in order to cut off the possibility of later competition from an equally effective product. At the end of six months the total cost for this effort has risen to $200,000.

Now the company draws up patent applications and files them. It sets up a pilot production line, which turns out to have a number of bugs in it. There are many more problems in producing large quantities of adhesive of reliable composition than in making small samples. Because of the scale of effort, modifications in the process become far more expensive than they were before. Quality control becomes a problem. Shelf life has to be examined. Use tests, which had already been conducted on a small scale, are now undertaken on a large-scale, long-term basis. Word of the development has gotten around, and it is decided to undertake a crash program in order to market the product before fall. A year and $600,000 have gone by.

By the time the first full production line has been set up, the marketing strategy has been settled, the sales force has been educated to the product, the use tests have been completed and evaluated, quality control techniques have been established, two years have elapsed and $1,200,000 has been spent. Now the company can market the product even though bugs in production, reliability, and quality still crop up. The product may or may not achieve the volume anticipated for it. The company may or may not make a profit.

The process described above is no longer and no more expensive than average for product and process development as they occur in medium to large firms. Some new products (a new acoustic amplifier, for example) have been invented, developed and brought to the point of marketing for as little as $5000. Other products – a new washing machine, a new electronic component, a color television process – may take over ten years and require between $10 and $20 millions before they are marketable. The figure of 'seven years from conception to market' is frequently cited. Regardless of scale, however, the shape of the development cost curve remains relatively constant:

Figure 5

Often, the first invention – the first demonstration of an effect – requires no more than a month or two and a few thousand dollars. As the company takes its first exploratory steps, the rise in the rate of expenditure is slow. As the company passes further check-points, the rate of expenditure increases. The total commitment grows in logarithmic fashion. Each new major commitment requires complementary commitments. Only when the product or process reaches commercialization does the curve begin to level off.

Then, in many businesses (consumer cleaning products, for example, or appliances), after a brief leveling off, as the company makes major commitments to advertising, marketing, and productive capacity, a step up to a new level of expenditure occurs. All of this precedes any real return to the company.

The S-curve for the cost of development takes on meaning when it is put in the context of the corporate investment game and the conversion of uncertainty to risk. Despite careful efforts to establish checkpoints beyond which development will not go without adequate justification, many companies find themselves having to make investment decisions on insufficient evidence in a general climate of uncertainty. As they begin to climb the slope of the S they rather quickly reach what appears to be a point of no return: the fat is in the fire. At this point, the executive vice-president can say to the president, 'We have put in so much, we may as well put in a little more in order to find out whether our investment is worth while' (to which the president can always reply, 'Don't throw good money after bad').

Occasionally, after the point of no return there comes a point where the mistake – if it is a mistake – is too big to admit. Large-scale developments of the kind undertaken by supercorporations or the military may proceed for months or years beyond the point where they should have stopped because of massive commitments to errors too frightening to reveal. Every corporate manager has experienced the difficulties of stopping questionable development projects once they are under way. They have their own momentum. In these cases the personal commitment of the people involved in the development, the apparent logic of investment and the fear of admitting failure all combine to keep the project in motion until it falls of its own weight.

These features of innovation derive greater force from our empirical knowledge that most new products and processes fail.

Efforts have been made to determine rate of failure. It has been estimated variously at between 7 to 1 and 20 to 1 and 40 to 1, depending largely on definitions. The farther back in the process of invention we go, the more overwhelming the rate of failure. In the absence of clear criteria of success or failure and of adequate statistics, it is not very useful to attempt a quantitative analysis. It is, at any rate, more accurate to say 'Almost nothing new works', than to say 'Most new developments succeed.' It must be added, moreover, that a general knowledge of the tendency of new products and processes to fail is present, to varying degrees, in the minds of those who undertake their development in corporations.

What we have been saying can be summarized as follows:

The process of innovation is one of converting uncertainty to risk. Corporate work goes into tailoring the messy process of invention to a form suitable for management's investment game. Questions concerning technical feasibility, markets and novelty, among others, must be answered on insufficient evidence, by individuals confronted with more information than they can handle. The inherent uncertainty of innovation rests in no small measure on the nonrational characteristics of invention.

The process takes time and costs money. The rate at which costs pile up makes small beginnings rapidly grow to major financial commitments, all before the uncertainties of development can be resolved. The company incurs the growing danger of yielding to the momentum of its own commitment.

From the point of view of corporate decision the process of innovation is essentially one of grappling with uncertainty, a process made more dangerous by growing pressures of time and money. The very resolution of uncertainties establishes commitments which make turning back more difficult, regardless of what future evidence may show.

The problem of innovation within the corporation is therefore a problem of decision in the face of continuing uncertainty. *A man must take leaps – not once at the beginning but many times throughout the process – in the face of uncertainty and on the basis of inadequate evidence.* The need for such leaps of decision grows out of the uncertainty inherent in the process. A company cannot escape it by careful planning or by gathering exhaustive data. The uncertainties resist resolution ahead of time. And the process of attempting to resolve them is itself a form of commitment, which plays its role in the cost curve and has its own momentum.

5 Peter V. Norden

Useful Tools for Project Management

P. V. Norden, 'Resource usage and network planning techniques', in B. V. Dean (ed.), *Operations Research in Research and Development*, Wiley, 1963, pp. 149–68.

Preface

The following article is an updated version of Chapter 5 of *Operations Research in Research and Development*, edited by Burton V. Dean (in 1963) under the title 'Resource usage and network planning techniques'. In retrospect, it is both gratifying and a little disconcerting to note that the concepts and principles set down seven years ago are still viable at this writing. One must conclude that the belief that there are useful underlying regularities in the research and development process is upheld by the continuing acceptance of both network planning and scheduling techniques, and the 'phase' or 'life-cycle' approach to funding and reviewing R & D projects.

On the other hand, little progress has been made in improving *a priori* estimating techniques in the R & D area, over what was known in the 1950–60 era. Such an indictment must be taken in context, however. By definition, almost, the environment and substance of R & D is uncertainty; and its mission, in the sense of entropy as expounded by Sir Stafford Beer, is to reduce uncertainty. This state of affairs alone precludes a fair comparison with other activities of, say, quantity production where statistical variance is so drastically reducible as to permit sophisticated quality control, and sharp estimating techniques based on standard costs and good historical data.

If we are still not appreciably better at making accurate *absolute* forecasts regarding cost and time requirements to achieve a given increment of technological advance, progress *has* been made in understanding the anatomy and physiology of R & D projects. This understanding enables us to talk about patterns of expected behavior (relative manpower and money and facilities requirements) of projects during various stages of their existence. In turn, this allows more sensitive tracking of these patterns, and to make judgements about 'project progress' whenever significant departure from the expected pattern is or is not observed.

Improvement in tools available to the manager for the planning and control of R & D projects is largely due to two forces in our economy:

1. An increasing inventory of quantitative techniques, (some heuristic, some analytic, and some via simulation), for scheduling, optimizing resource allocation, and forecasting.
2. Significant progress in computing technology, notably data-base oriented systems, time-shared remote-access operation, and graphic and interactive terminals and displays.

Network techniques (PERT, CPM) and life-cycle methods are examples of the former. The latter permit increasingly rapid detection of plan-*v.*-actual deviation and appropriate management corrective action. Further, in a very real sense, does the availability of powerful remote-access computing bring about the reduction-to-practice of the conceptual tools.

One further observation: the intervening years have taught us that the search for a single project management system, to serve the needs of all levels and functions of management, from birth to death of a project, is fruitless. Attempts to develop such a monolithic multi-purpose system, particularly in the form of stand-alone computer programs, have generally resulted in complex monstrosities. One wag has compared these to a 'Cadillac with half-tracks'. This observation should not be confused with the design of compatible data bases, which very usefully serve as the standardized connective tissue of loose aggregates of computer program routines.

The complete project manager's tool-kit will probably take the form of a compatible collection of such routines, each constituting a model of some aspect of the project's operation, and of interest to a particular manager. For example, the following categories of aspects of projects and their institutional environments appear to require different techniques:

1. Stage, or time in the project life. Significantly different decisions need to be made before a project is given the go-ahead nod, and once it is under way. Further, information needs, involving the kind, detail, and frequency of data, differ in the early, mid, and late stages of a project.

2. Management hierarchy level. Aggregated figures, broad trends, and early warning on the *trouble no trouble* spectrum are most valuable to higher levels of management. First-line management needs access to these, but *in addition* requires anatomical detail, checklists of things to be done 'next week', and pinpoint data for short time horizons for dispatching and tactical correction actions.

In this context, the techniques discussed below have the following relevance:

1. The life-cycle method is useful to middle and higher level management to answer questions of 'How long will the project take?' and 'How much technical manpower will be required.' It is applicable in the before go-ahead stage, and for monitoring once under way. Life-cycle systematically makes good use of accumulated historical data, and can accommodate plausible estimates for a 'cold start'. It is less useful at the working level for day-to-day managing.

2. The usefulness of critical path techniques spans a wide spectrum, being applicable for higher-level planning with networks of a comparably high level of summarization, but has its greatest utility at lower level management of detail tasks. A unique broad-spectrum capability is the ability to integrate many detail networks into a composite, thus making visible some non-obvious consequences of intricate interdependencies of the subnets. Some progress has also been made in assistance with the allocation of resources to jobs on the PERT net.

3. Finally, some advances have been made in the area of assessing risk of failure to meet schedule. A postscript has been added to the article, describing how certain risk curves can be used for orderly recording of consensus regarding the status of a project, and tracking departures from the original risk assessment.

The introduction of PERT made it practical, perhaps for the first time, to record complex, involved human affairs in a systematic and comprehensible way. It is, therefore, not surprising that managers of R & D projects are developing a taste for exploring whether this capability cannot be extended to other variables of interest, such as the addition of resource-planning capability to PERT-type networks (Voress, Houser, and Marsh, 1961). This chapter discusses some experimental work done in this area at the I.B.M. Development Laboratory, Poughkeepsie, New York, for potential use by R & D managers.

We will present certain resource-utilization patterns on engineering projects and their relation to network schedules.

The study of resource consumption and network schedules involves the conduct of engineering work, internal organization, and book-keeping, so that a number of companies have understandably decided that publication is against their best proprietary interest, even though they may be actively working in this area. Therefore, we may not be aware of comparable work as well as work that has gone beyond that being discussed herein.

In addition, we will presume the following:

1. We will assume that the reader is familiar with the Navy's PERT system in its broad forms (Malcolm *et al.*, 1959; PERT, 1958). Therefore, no attempt will be made to give a description of network or critical-path-type scheduling and planning techniques.

2. We will deal only with our experience with applied R & D projects. This will eliminate fundamental or basic research and quantity manufacturing activities from our discussion.

3. We will confine our attention to homogeneous projects. Primarily, we wish to avoid confusion with what has been variously termed 'multiproject overlay', multicontract, 'total program' or facility-wide models. These depict the universe in accordance with a somewhat different scheme of classification.

The Project as a Conceptual Entity

Division or classification is a major conceptual problem in its own right, which has been the subject of serious inquiry. Unfortunately, these problems are beyond the scope of this chapter.

Sir Stafford Beer (Eckman, 1961), states, among other points, that a project is a natural base for cohesive activity and suggests that a manager who bears the title of 'Sales Manager' might be more appropriately titled 'Manager of Project Number Six' in the sense that this encompasses a more meaningful set of activities, over which he exercises judgment, choice, or control.

In an earlier article (Norden, 1960), a tentative definition of a project, which will be used in this chapter, was proposed:

> A development project is a finite sequence of purposeful, temporally ordered activities, operating on a homogeneous set of problem elements, to meet a specified set of objectives representing an increment of technological advance.

In the same article, homogeneity was defined thus:

> We will call a task homogeneous if it is composed of elements, each of which has at least one technological interdependence point in common with another element during their life-cycles.

When we formulated this definition, we had not had experience with PERT, but in retrospect, the compatibility of concept becomes apparent. The primary characteristic is that of connectedness, and this serves to distinguish a development project from the environment in which that project is embedded.

By this definition, a set of activities will comprise a project if the activities are all connected by a set of technological interdependencies, and the boundary of the project will be wherever further interdependence lines cannot be drawn logically. In terms of PERT practice, a project ceases with the terminal points which knowledgeable people have elected to define in the network diagram. Experience indicates that this is not a rigorous relationship, but one that is usefully approximated in practice. In a similar sense, the meteorologists' 'air parcel' cannot be demonstrably carved out of the gaseous envelope of the earth with finite boundaries, but has the property that certain variables of interest, such as temperature, pressure, or closure of wind patterns are more cohesive within than outside. The project, therefore, is a convenient conceptualization for study, although we know that every project has interactions and connexions with suppliers, contractors, and contract administrators, generally outside a limited geographic perimeter, but these become increasingly tenuous beyond the 'entity' of the project. We will, therefore, follow the operational approach of saying that the project boundaries exist where knowledgeable people have determined to cease drawing further PERT-type activity lines.

Networks and Resources

Let us now consider the network schedule of some homogeneous project. This is a partially ordered set of events and activities and constitutes a topological system in the time domain.

This is an important point. The *network* representation of a development project or comparable enterprise is meaningful only in the dimension of time. That is, the sequences of activities are well-identified time precedence relationships. When experience with the time aspect of critical-path-type schedules had become familiar, it became logical to investigate the addition of resources to such a time network by using the PERT concept.

Initially, this implied associating three estimates for money, manpower, or other variable of interest with each line of the network, in the 'optimistic, realistic, and pessimistic' fashion of PERT. Presumably, this would yield the capability of computing probabilities of spending no more or no less than a certain total number of dollars, with the same confidence now possible for elapsed time. Since resource usage and the time to do an activity are not independent, this scheme is equivalent to asking for nine points on a micro-economic response surface by specifying time and money *pairs*. A comparable relationship, but one which also has some intrinsic differences, is utilized in the Mauchly Associates' (Kelley and Walker, 1959a and b) critical-path algorithm and in Fulkerson's algorithm for network optimization (Fulkerson, 1961), which was used by General Motors in the SHARE 1188 computer program for the I.B.M. 704 and has also been programmed for a number of other machines by several computer manufacturers. More recently, the computer industry has produced pre-packaged application programs, such as I.B.M.'s Project Management System (P.M.S.) for the I.B.M. System/360. This program contains four major 'processors', or sections, which can be combined in elaborate ways: a *network processor*, for processing PERT nets in the time domain, both on an individual subnet basis and as an integrated composite network; a *cost processor* for associating money, manpower, and other resources with the jobs represented on the time nets, and performing sundry accounting summarizations thereon; a *resource allocation processor* for distributing available resources among the activities that need them, so as to minimize period-to-period fluctuations, not exceed ceilings of availability, and to minimize schedule or budget overruns under stated management priority policies; and a *report processor* for formating the information that has been operated on in the other three sections, in an understandable manner. These programs go a long way toward enhancing the project manager's tool-kit, discussed in the preface, but still fall short of a full-fledged, algorithmic economic time/cost 'trade-off' function.

In R & D work, this still poses a number of practical problems. We can use a type of linear approximation between a crash execution time and a normal execution time, or a piecewise linear approximation, whenever we know more points in the interval, provided that we are dealing with activities which are well-enough defined to constitute jobs with 'standard content'. Lacking these, it can become very difficult to accumulate the necessary historical cost records which would give us confidence that the stipulated pairings of cost and time for the crash, normal, or intermediate situations are indeed true, or permit us to assign some equivalent 'trade-off' function. Furthermore, it has been asserted that piecewise linear or linear interpolation may be totally meaningless in many situations since these functional pairings may be realizable only at certain discrete points.

Another major question in intertwining resource information and PERT networks is the problem of identifying or defining the work content of a given line of activity. To be sure, this may simply be a matter of the mechanics of executing the PERT or the PERT-cost concept in an actual situation. However, it seems that much of the preoccupation with the PERT-cost question is one of 'how can we fit PERT schedule data, and information derived from PERT networks, into our present accounting structure?'

To the extent that we have encountered this problem, it led to a realization of the inappropriateness of many existing accounting definitions; i.e. what work is to be charged to a work order, shop order, project number, job number, etc. Rarely have mercantile charts of accounts provided sensible groups which would enable us to achieve rational acts of classification. Happily, in our laboratory we have made definite strides in reconciling a rational, operationally manageable basis of engineering work activities with the reporting requirements for corporate profit-and-loss and Internal Revenue Service accounting needs.

This, by the way, is an interesting development from the Operations Research point of view. The PERT network, drawn for a homogeneous project as discussed in the foregoing, *does* represent a cohesive entity of work. Management agencies now wish to associate resource–consumption variables with this network but want to fit them into categories which have no meaningful relation to the intrinsic work content of the project and the net. We are, in a sense, asked to map a set of variables from one domain into others without being able to define a workable transformation function. We think that this is an important methods and procedures problem, and since management has many obligations in the conventional accounting and finance realms, ways must be found to find these mapping functions, procedures, and computer programs to bring the two requirements into harmony. We only hope that we can meet somewhere in the middle. A promising compromise is to associate bundles or subsets of activities with the job orders at the ledger level.

However, we should not deceive ourselves into thinking that, if we make PERT network information compatible with total management operating systems or existing accounting and reporting structures in the sense of integrated data processing, we are dealing *structurally* with an extension of the network scheduling concept.

The former is a procedural problem. The distinction seems to lie in this area: the network, as we have seen, is a topological construct. Resource consumption, in our estimation, simply does not have an intrinsic mathematical structure in the class of directed graphs or networks. Rather, we are dealing with a set of resource–consumption vectors associated with each line of activity or sets of such lines. Only the event-sequencing is a partial ordering in the time domain. This point of view can free us from what might be a mental strait jacket, namely, to try to form money networks, manpower networks, facility networks, reliability networks, etc.

There are, of course, very valid approaches (sometimes facilitated by suitable computer programs) which try to compare resource–consumption forecasts associated with PERT nets with available resource levels, and identify critical shortages or bottlenecks (McGee and Markarian, 1961), and the P.M.S. program battery referred to earlier. Here again, however, we are dealing with resource consumption 'bills of goods' associated with a time-domain network or networks.

This may have sounded negative; however, we gain a great deal when we think of mathematical models of resource consumption *associated with* time networks. What we gain is access to the vast and rich field of mathematical models. For the first time, PERT-type networks can give us an insight into the real internal anatomy of engineering development that underlies the expenditure patterns.

In the next section this point will be illustrated with one model which we have reduced to practice in the last year and a half. We have called it the life-cycle method of project planning and control. It is a mathematical model relating manpower usage on engineering development and similar projects to calendar time.

The Life-Cycle Manpower Model

Life-cycle is concerned with manpower curves encountered on applied research and engineering development programs. These curves depict manpower utilization on R & D projects between the start and completion of all work necessary to comply with specified and agreed-upon program objectives as a function of calendar time.

Let us recall that a project can be (and generally is) broken down into smaller parts or elements. What these elements are and how they are to be scheduled relative to each other in time, as for example in the network sense of the PERT programs, is determined by the technological content and structure of the project. The series-parallel mix of topological relationships in such a schedule implies that some *portion* of the needed work on some elements cannot be begun unless some other *portion* of work on some other element has previously been completed. This is what we designate as 'technological interdependence among elements'. If no such interdependence relationships existed in a given project, all elements could be started and worked through in parallel, without regard for any other elements. If there were m such elements, we would view these as m independent projects.

Conventional usage of the term 'project' could lead to confusion in this context. People use the words 'project' and 'program' interchangeably to designate a grouping of tasks by some scheme of classification, such as common subject matter, under a common manager, supported by a common appropriation, etc. Although such groupings exhibit logical consistency, they do not necessarily require *connectedness*. In the foregoing, we have equated connectedness with homogeneity, and the latter with the notion of 'project'. Thus, to talk about a 'homogeneous project' is admittedly redundant but is here used for emphasis. We wish to address ourselves to homogeneous projects. A project is homogeneous if it is composed of a set of elements such that each has at least one technological interdependence tie to at least one other element at some time during the life of the project. Attention is now called to the following aspects of the homogeneous project:

1. Finite problem sets.
2. Operations: activities; work; what people *do*.
3. Operands: elements; problems; what they *do it to*.
4. Purposes: objectives; *why* they do what they do.

The last point is important in the life-cycle model. The purposes change throughout the life of a project, and these changes characterize the effort cycles which will be discussed later.

Research studies have indicated (Norden, 1961) that there are regular patterns of manpower buildup and phase-out in complex projects. These patterns have been described by a number of mathematical functions, generally in the family of exponential, gamma, beta, or logistic curves, by several investigators (Norden and Bakshi, 1960). In the life-cycle model, curves are fitted to a small number of successive 'cycles' of work which occur during the life of a project. The cycles do not depend on the nature or work content of the project but seem to be a function of the way groups of engineers and scientists tackle complex technological development problems. Each cycle can be described by a comparatively simple equation:

$$y' = 2Kate^{-at^2}$$

where y' = manpower utilized each time period,
K = total cumulative manpower utilized by the end of the project,
a = shape parameter (governing time to peak manpower),
t = elapsed time from start of *cycle* (Figure 1).

Thus the cycles can be represented by a series of curves from the same family, relating manpower used each month to elapsed calendar time, and differing only in relative size and proportions. The single parameter governing the shape of the curves can be

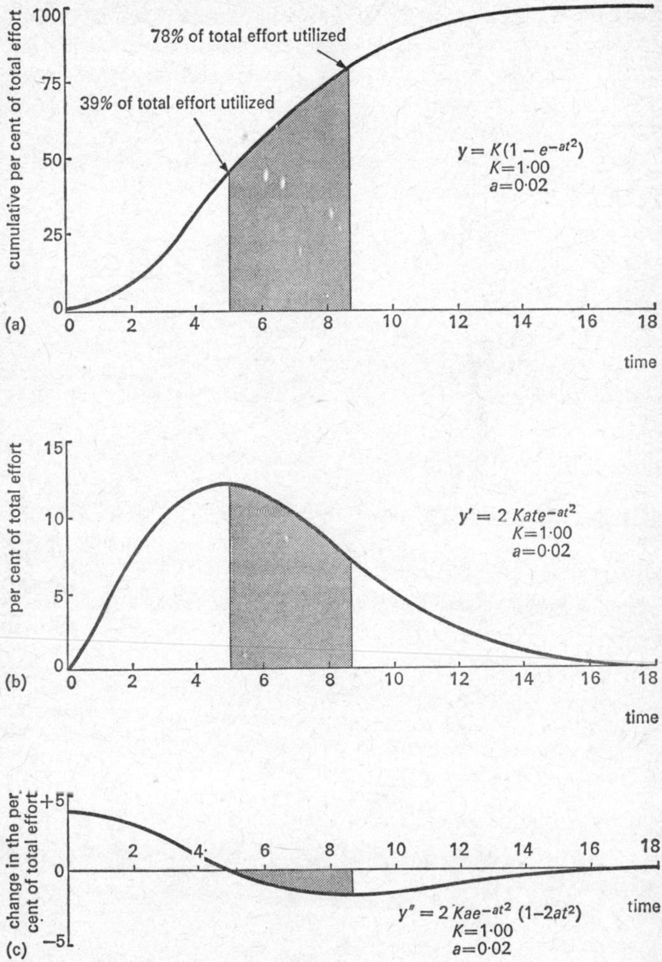

Figure 1 Manpower utilization curve: (a) cumulative manpower utilization (b) current manpower utilization, and (c) the change in monthly manpower utilized

thought of as a measure of the 'importance' of the project (Figure 2). Sharply peaked manpower buildups correspond to crash projects, while shallower curves are associated with stretched-out projects.

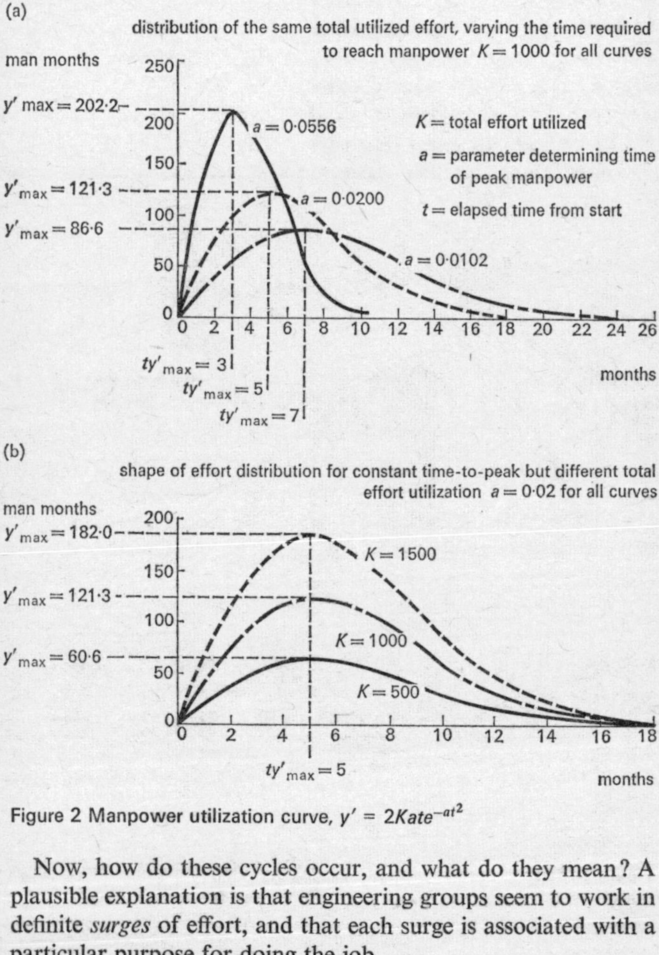

Figure 2 Manpower utilization curve, $y' = 2Kate^{-at^2}$

Now, how do these cycles occur, and what do they mean? A plausible explanation is that engineering groups seem to work in definite *surges* of effort, and that each surge is associated with a particular purpose for doing the job.

In the I.B.M. Development Laboratory in Poughkeepsie, New

York, the succession of purposes with which we work on projects generally are: planning, designing, building, and testing a prototype, engineering activities associated with release of the product to plant, occasional redesigning, and a small number of cycles for product support and cost reduction (Figure 3). Sometimes this sequence is preceded by an exploratory cycle. The whole

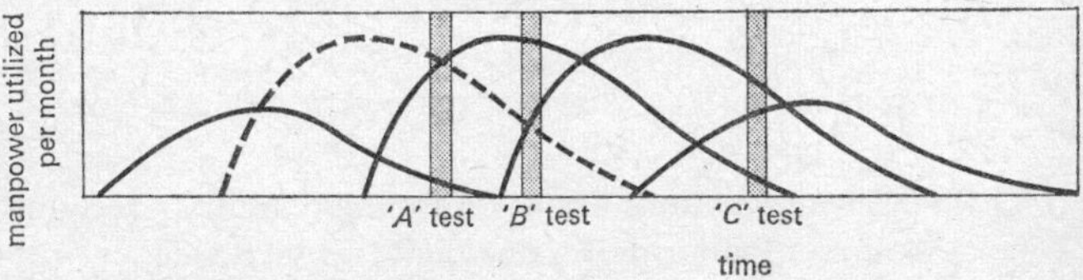

planning cycle
- define output (technical objectives)
- investigate problem areas
- set specs. to guide subsequent design

design cycle
- design, build and test data flow models
- design the prototype, order hardware
- produce all documents needed to build prototype

model cycle
- build the prototype
- test and debug the prototype
- redesign of unreleased parts to meet specs. or to obtain 'B' test approval
- release documents necessary to manufacture the product

release cycle

all engineering lab activity on released parts
- to enable manufacturing to produce first item
- to remove flaws in original design
- to update the product's technology

product support cycles

all subsequent engineering activity
- factory service
- cost reduction
- maintenance reduction

Figure 3 Typical manpower pattern of an engineering project

sequence corresponds to one often encountered in human endeavor:

Explore	Establish feasibility.
Plan	What shall the product be like?
Decide	How shall it be designed in detail?
Do	Build it according to the plan.
Refine	Correct things which turned out differently from what was expected.
Improve	Incorporate knowledge gained in the foregoing.

These cycle designations are part of the laboratory's account-code structure in the form of a one-digit 'activity code'.

The mathematical model of project manpower consists of the equation for each cycle plus a linking function which specifies the relative sizes and durations of the cycles and their lags or spacing in calendar time. The linking relationships have been encouragingly stable over a wide range of projects and a number of years. This fact makes it possible to develop projections of manpower and time requirements for comparable projects, given a few actual points on the early cycles. In addition, early warning of significant departures from current schedules can be obtained by monitoring the actual progress of a project against a prior projection is, of course, a classical control function of management. This is sufficiently important to warrant slightly more elaboration here.

Let us recall the curve of a single work cycle. This curve has many features which make it ideal to work with computationally. These features form the basis for the simplest method of estimating the coefficients of the manpower utilization curve. They enable one to arrive at quick results of considerable accuracy. Four of them are of practical value for rapid analysis, and many may be derived by the use of differential calculus:

1. The coefficient a determines the month in which manpower utilization is greatest. One can calculate this month by the formula

$$t_0 = \left(\frac{1}{2a}\right)^{\frac{1}{2}}.$$

2. Conversely, if one knows the month in which manpower utilization was or will be the greatest, he can calculate the coefficient a:

$$a = 2t_0^2$$

3. If one knows the month in which manpower utilization has reached or will reach a peak, and if one knows or can estimate the manpower level in that month, one can calculate the value of K, the total manpower required in the cycle, by substituting for a as determined above, and solving:

$$K = e^{\frac{1}{2}} t_0 y_0.$$

4. The manpower utilization curve has a point of inflection at a point at which the decrease in monthly utilized manpower slows down in the descending portion of the curve,

$$t_1 = \left(\frac{3}{2a}\right)^{\frac{1}{2}}.$$

This point is useful in setting milestones for a project. When the cycle has passed this point, the work should be in the 'cleanup' stages.

We compute the manpower cycle curves either by using the foregoing relationships or (particularly in projects which are already under way at the time at which the first life-cycle analysis is prepared) by statistical curve-fitting techniques which we will not discuss in detail here. Then we establish control limits at ± 10 per cent of the predicted monthly manpower. This control interval was selected arbitrarily, and will be modified if it turns out to be too wide (failure to detect serious manpower overruns) or too narrow (too many 'nuisance calls' to the project manager), after we have more experience. So far, the interval appears to be a reasonable start. If the actual manpower usage of a project goes outside the control interval (Figure 4a), a new projection is made (Figure 4b).

This is discussed with the project manager and, if he concurs, his manpower forecast is revised. Since an overrun in an early cycle cascades through all the subsequent cycles, this 'early warning' potential of life-cycle is applicable.

One note of caution must be sounded. The linking functions

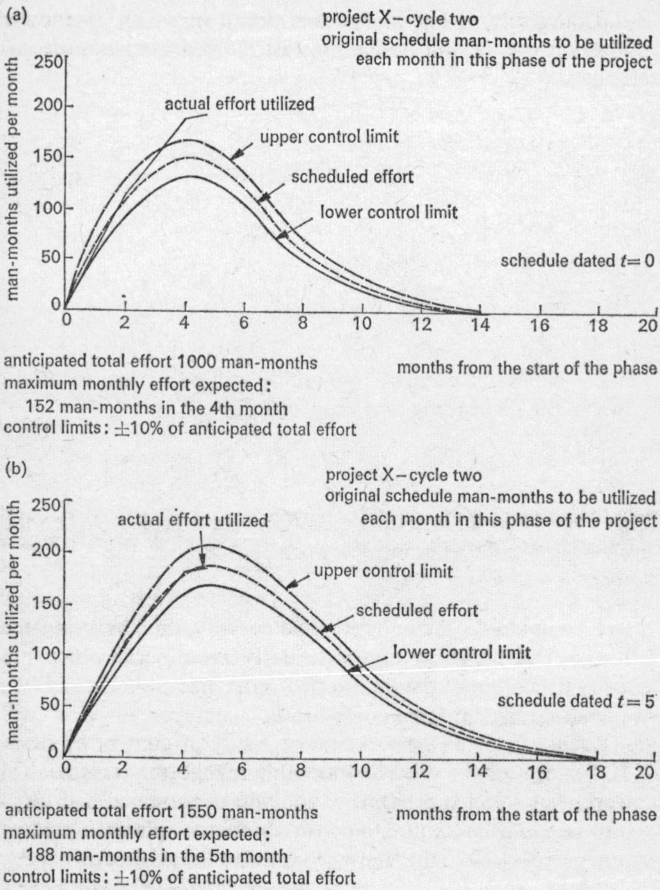

Figure 4 Life-cycle method of project control – an illustration of the early warning technique

(ratios of manpower and duration relating adjacent cycles) exert considerable leverage on the total project manpower forecast. We make a point of emphasizing to our management that our project engineers are not irretrievably 'locked in' on the laboratory's historical performance. Quite to the contrary, as improved ways are found of conducting the technical work, our most fruitful experience to date will become the new reference patterns for future planning.

It is in this exercise of continuing improvement that the simultaneous use of life-cycle and PERT-type planning complement each other well. The curves are statistical aggregates. The network gives anatomic detail. Prior to the advent of PERT, we could only point out to a program manager that 'in the third quarter two years hence, your estimated manpower is only about one half of what history indicates you will probably need'.

We could not tell him *what activities* he would be working on at that time except in the broadest sense. PERT, of course, can almost pinpoint the particular jobs on which he anticipates to be working during the time period in question. This enables him to confirm or modify his manpower estimates well in advance.

In the light of the foregoing, the point made in the section entitled 'Networks and Resources' can be clarified. In most schemes currently called the PERT–cost concept, we are merely labelling each line of a PERT network with the manpower or other resources which we estimate it will consume. Even at the level of algorithmic abstraction presented in other approaches (Freeman, 1960), it is assumed that the trade-off of resources and time is *known* and can be supplied at the time at which the inputs for a computer run are prepared.

Even if we ignore for the moment that, at the level of detail of one PERT activity line, it is often very difficult to estimate two points for a linear trade-off, we would still be populating the network with guesses. Life-cycle, by supplying stable patterns of larger work aggregates, provides a measure comparable to the behaviour of a gas in a closed container: mankind has found the notions of temperature and pressure of considerable practical value without the need for knowing the precise Brownian motions of each molecule.

This leads us back to the fact that project planning is best done at the detail PERT level with an activity orientation, whereas evaluation and control are profitable exercised at a more aggregate level with a functional orientation. We have found that the life-cycle model provides an orderly summary, crystallized out of past experience, at a functional level. It gives us the capability of conveniently using historical experience to shed light on the future.

Theory of Project Manpower Patterns

Up to the present time, we have applied the method outlined above to about twenty projects with encouraging success. The inherent problems of labor-claiming and accounting practices limit the accuracy of the 'actual' data reported to us through conventional administrative channels. In addition, the inevitable time lags associated with phasing personnel into and out of projects may distort the canonical shape of the curves slightly. This prevents us, for example, from questioning an actual manpower charge of seventy-five men in a given month, when the life-cycle projection would have indicated seventy. Nonetheless, the basic pattern has been observed in a number of development engineering situations, both inside and outside of our laboratory.

We have developed the following theoretical model to explain the observations. We assert that an engineering development project can be viewed usefully as a set of unsolved problems. This unsolved problem space is exhausted in the presence of human effort and creativity in a deliberate, purposeful group-problem-solving setting. The operation of exhausting the problem space involves many activities which need not be explicitly ordered in time. However, there is a partitioning of sets of these activities. This divides the unsolved problem space into a group of subspaces, which correspond to the purposes with which the problem-solving operation is concerned at various stages in the life of a project. In practice, the exhaustion operation is the design decision-making operation. In this context, the several problem subsets, each labeled by its *raison d'être*, represents a stagewise conversion of problems into solutions.

If we make the following assumptions concerning *each* such subset, the effort-utilization patterns actually observed can be constructed:

1. The number of problems in the subset is finite, albeit unknown.

2. Human problem-solving effort constitutes an environment for the unsolved problem set and makes an impact on the problems in the set.

3. Information gathering, gestation, identification of alternatives open for choice, and deliberation consume varying amounts of time. Any planning or design decision, made as a result of such deliberations, represents an event which causes one unsolved problem to convert into a solved problem, thereby removing it from the unsolved problem space. If we assume the occurrence of these events to be independent and random, then, by the Poisson model (to be discussed later), an exponential distribution of interevent times is a reasonable assertion. It is further possible to postulate a conditional probability function for the event that a problem will be *permanently* removed from the unsolved problem space, given that an identification event has occurred. Stated in another way, we postulate a conditional solution, given a certain distribution of insightful situations.

4. The parameter of the decision-event distribution is a compound, implicit function of skill levels of problem solvers, level of exertion, administrative actions, and other random manifestations of the problem-solving situation, and interaction with the environment.

5. The number of people working in the group at any given time is approximately proportional to the number of problems 'ripe' for solution at that time. This leads to an interesting interpretation: The problem solvers act in a dichotomous manner as *catalysts* or products of the problem-solving process as much as *agents* who cause problems to be solved.

The observed patterns can be shown to follow from the above assumptions. The model leads to a succession of surges of effort or cycles throughout the life of the project. The plausibility of the above process was underscored by an excellent study by D. L. Marples at Cambridge (1961). He postulates that engineering design decisions have a treelike structure in which technical choices are made at each mode. Subject to some technical criterion, one of a number of admissible alternatives is selected, symbolically representing a branch which leads to the next (problem) node.

Consider now a reliability model due to Epstein (1958). He discusses a set of independent devices under test, subject to some environment E, which is a random process. Suppose this environment generates shocks, or some 'peaks' which are destructive to the devices in question. It is then reasonable to assert that such peaks (thermal shock, extreme vibrations, etc.) are distributed in a Poisson manner with some (rate) parameter.

Let T be a random variable associated with the time interval between successive peaks. Then

$$\Pr(T > t) = \Pr[\text{no event occurs in the interval } (0, t)], \quad \mathbf{1}$$

where $t = 0$ is the time when the most recent event occurred.

Then, from the Poisson assumption

$$\Pr(T > t) = e^{-\lambda t} \quad \mathbf{2}$$

and

$$\Pr(T \leqslant t) = 1 - e^{-\lambda t}. \quad \mathbf{3}$$

The probability density function associated with **3** is

$$f(t) = \lambda e^{-\lambda t}. \quad \mathbf{4}$$

This would describe the failure distribution in an all-or-none situation: Given that a destructive peak has occurred, one device will fail with certainty. The number of items remaining and the (average) number failing per time period will also follow exponential functions.

Now suppose that the above situations were governed by a *conditional* probability of failure rule: Given that a peak has occurred at t, the probability of failure is some constant p $(0 \leqslant p \leqslant 1)$. Then it can be shown that

$$\Pr(T \leqslant t) = 1 - e^{-p\lambda t} \quad \mathbf{5}$$

and

$$f(t) = p\lambda e^{-p\lambda t}. \quad \mathbf{6}$$

This would result in a time-invariant conditional failure probability. Whenever a shock occurs, a device will fail with fixed probability. The number of items remaining and number of items failing per time interval are again exponential functions.

Now consider the conditional probability of failure to be time dependent. Then it can be demonstrated that if the conditional probability of failure is some function $p(t)$:

$$\Pr(T > t) = \exp\left[-\lambda \int_0^t p(r)\,dr\right], \qquad 7$$

whence $$\Pr(T \leqslant t) = 1 - \exp\left[-\lambda \int_0^t p(r)\,dr\right] \qquad 8$$

and $$f(t) = \lambda p(t) \exp\left[-\lambda \int_0^t p(r)\,dr\right]. \qquad 9$$

This results in the class of Weibull distributions, well known in reliability work. The physical interpretation here would be that the probability of devices succumbing to destructive shocks is changing with time. In our case, most of the data we have observed (the life-cycle curves)

$$p(t) = at. \qquad 10$$

This could imply that our engineers are learning to solve problems with an effectiveness which is increasing linearly during each cycle. When one considers that familiarity with the problems at hand can lead to greater understanding and sureness, this result is not implausible. We are experimenting with other learning curves:

$$p(t) = at^{-b} \qquad 11$$

of which our present model is a special case. Suitable choice of parameters appears to explain a number of other manpower-utilization patterns found in the literature.

Conclusion

We have indicated that by suitable analogy to Epstein's work, problem solving of engineering development groups may be dependent on:

1. A group idea-generation rate or the rate at which sets of potential solutions are formulated by the group for each previously identified problem.

2. A conditional probability function governing, once a set of potential problem solutions has been identified as being (by application of suitable criteria) a feasible and acceptable solution which removes the problem permanently for the unsolved problem space.

An increase in the time rate of change of (2) is a type of group learning – a time-dependent process which increases the probability that, with time, Marples' 'branch bundle of alternatives', once identified, will contain an ultimately acceptable solution.

One last word on the idea-generation rate. Lorge and Solomon (1955) suggested that problem-solving effectiveness of a group is a function of the capabilities of its most competent members. Suppose that in an R & D project, this is the project engineer or some key man. He is the man who sets the pace for the program, or at least has the insight to attract a highly competent key member who in turn sets the pace. In either case, then, there is one or a small number of upper limit persons whose insight-producing rate determines the rate parameter (λ) or the exploitation-effectiveness parameter (a) of our formula. We would, therefore, assume that these people have the capabilities either to make their insights faster and/or to produce potential solutions which are closer to being right and can, therefore, be finalized more quickly. Or, in the final analysis, they may simply produce so many more insights that there is greater possibility that one good avenue of potential among them is a good one or a fruitful one. Or, again, it may be possible to regard the project group as being composed of two types of people – idea identifiers and idea evaluators. The latter could be thought of as applying feasibility and optimality criteria to the potential solutions developed by the former. It is not necessary for these two functions to be executed by separate (specialized) individuals, but the same individuals could be both creating and applying value judgments at different times or even in rapid 'cycling' alternation. In the last analysis, the life-cycle model suggests that we are dealing with two processes – generation and exploitation. Both may become limiting in their effect on project progress. Both are, as suggested in **11** at least for the exploitation factor, subject to improvement by human learning.

Finally, the staffing of engineering projects is not inherently a matter of wide trade-offs of time and manpower. Rather, they seem to embody a two-stage process in which the limiting condition is the problem-identification rate or 'insight environment' of the unsolved problem space. This implies that one cannot indefinitely add people and get the job done faster. Intuitively, we know that this is so, but the current model leads us to the conclusion that the limiting factor is the rate at which ideas or insights can be generated, and that the rate is not widely affected, if at all, by the number of men on the job, but rather by some capability level of the group.

Therefore, within one cycle, for all practical purposes, the parameter of the (exponential) 'insight distribution' remains constant over considerable periods of time. The fact that we can add people periodically as more work pieces are identified is a derived effect.

As a matter of fact, to decide where to 'saw off' a given development program and to say that 'this is what we will produce rather than to continue development towards higher performance' is an implicit understanding that problems which are unearthed after a certain point in the incremental advance of technology would result in a parameter of a different value. In other words, we stop a given program when a significant advance in the direction of higher performance requirements becomes or is recognized to be much harder or more difficult. This is an intriguing speculation because it implies that, in some sense, the parameter of the 'Eureka' distribution is an indirect measure of the difficulty level of the problems at hand.

Considerable further study is evidently needed, but the combined tools of network analysis and mathematical models are opening promising avenues of attack.

Postscript: Risk Curves

Since the writing of the foregoing, it became obvious that the problem of project management, like that of most management, was to find an acceptable balance among time, cost, and performance. In the R & D area, PERT, resource allocation processors, and life-cycle methods address the first two in a rudimentary fashion. The harried project manager most often uses performance as his last resort escape, if he is really 'boxed in' on the other two factors by tight management. That is, if he is tightly held to budgets and delivery schedules, he may try to sell slightly derated performance (shorter than planned cruising radius of an aircraft; slower than planned computer memory; heavier or larger than planned space vehicle configurations, etc.)

The vast variety of subject matter and products which are, at one time or another, in research or development precludes glib reduction to formula of universally relevant performance measures. In this area, as in any others in which valid quantification is either absent or inadequate, projects are managed by the subjective judgment or feel of experienced professionals. Whatever cannot be explicitly formulated is implicitly intuited. In recent years we have used 'risk curves' as a device for orderly recording of expert opinion regarding the joint effect of time, cost, and compliance with specifications. These curves can also track changes of opinion or outlook from a project team's original position.

These curves appear as in Figure 5. In this particular case (the data have been disguised for proprietary reasons, but the relationships among relevant factors have been preserved) the chief engineer of the project and the key technical people on his team made all the estimates. Figure 5 depicts these estimates as of July 1967. At this time the project had been in the preliminary planning stages for about two months, specifications had been roughed out, and two fundamentally different engineering approaches, each of which would meet the technical specs, had emerged. The first (plan or approach A on the exhibits) was predominantly a hardware solution; the second (plan B) was software oriented. It should be noted that in this case the same development group would be assigned to do the job, irrespective of which approach would ultimately be chosen, and this was the group that drew the risk curves for themselves. The group contained both software and hardware experts, and we therefore assume that there was no built-in bias to 'stack the deck' in favor of either proposal.

The curves plot subjective estimates of the probability of failing to meet the total project commitment (delivery date, full compliance with all specification items, and within authorized budgets) as it exists at the time of estimating. The estimator takes into consideration all that he knows about the number, types, and ability of people he will be able to draw on, any delays in hiring additional people, availability of computer time when he is likely to need it, the state-of-the-art of the technology involved,

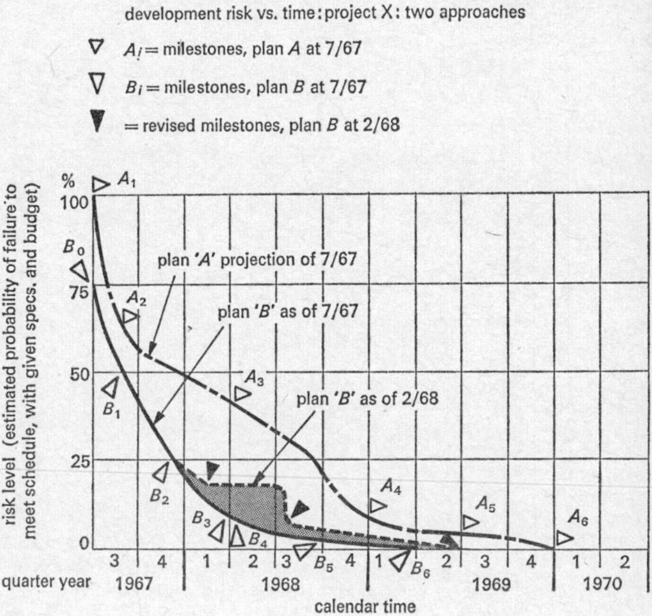

Figure 5

how much inventing may still have to be done, etc. In addition, a series of milestones is established. These generally consist of checkpoints of accomplishment, or standardized review points. The better these milestones have been institutionalized (i.e. the better their meaning is agreed upon and understood by everyone in the organization), the more reliable will the risk assessment be. A risk level is associated with each milestone, as well as an expected date of occurrence. The risk curve drawn through these milestones, then, has the following meaning: If milestone *i* actually is accomplished at time t_i, then the likelihood of *not*

Table 1

Project X: Analysis as of July 1967

Milestones			*Approach A*			*Approach B*		
A_i	B_i	*Meaning*	*Expected date*	*Risk level (%)*	*Cum. \$ × 10^5*	*Expected date*	*Risk level (%)*	*Cum. \$ × 10^5*
A_1	B_0	Orig. start	6/1/67	100	0	6/1/67	75	0
A_2	B_1	First review	9/1/67	60	1	9/1/67	50	1
—	B_2	Architecture CM	—	—	—	11/1/67	25	3
—	B_3	Feas. test entry	—	—	—	4/1/68	10	7
A_3	B_4	Announcement	5/1/68	40	10	5/1/68	5	8
A_4	—	Key software CM	1/1/69	10	18	—	—	—
A_5	B_5	Syst. test entry	7/1/69	5	24	10/1/68	3	13
A_6	B_6	Delivery	12/31/69	0	27	4/1/69	0	18

meeting the final commitment is at risk level R_t ($0 \leqslant R_t \leqslant 1$). A risk level close to zero would apply to 'sure things'; a risk level close to 1 (or 100 per cent) would imply the need for nearly impossible technological breakthroughs.

Generally, risk should decrease rather smoothly throughout the project's life. Whenever plateaus or bumps appear in the curve, it indicates concern on the part of the engineers that some key technical problem may be difficult to solve. Projects well within the state of the art would be expected to have rapidly dropping risk curves. More innovative projects tend to show shallower risk reduction and plateauing.

The data plotted in Figure 5 are taken from Table 1. It is evident that as of July 1967, plan A was thought to have uniformly higher risk, as well as cost $900,000 more and take nine months longer to complete (see Figure 6) than plan B. Since management was persuaded of the validity of these opinions, plan B was approved and embarked upon.

One of the advantages of risk curves is that they are relatively easy to construct. Of course, like any subjective technique, they are open to criticism that the resulting function cannot be empirically verified. But projects are managed subjectively anyway, and techniques such as the RAND Corporation's DELPHI, method (Helmer, 1967; Helmer and Rescher, 1959; Delkey and Helmer, 1963) are gaining increasing attention. It is possible to subject an entire risk curve, or selected milestones to the DELPHI, approach, in order to achieve convergence on consensus in a systematic and orderly manner.

Once information such as that in Table 1 and Figure 5 has been generated, computations such as those of Table 2 can be made. This shows that under plan A there would still be a 40 per cent chance of not meeting final schedule, at the time of announcement of the product. In the case of plan B, the risk is only 5 per cent. Further, in the crucial 'architectural' phase between first review and announcement, B will only be spending $20,000 per percentage point risk reduction, to A's $45,000; and B has the 'luxury' of spending nearly ten weeks and $200,000 per point of risk reduction, for forty-eight weeks, where A only has 2·2 weeks and $42,500 per point but requires eighty-seven weeks, in the final reduction-to-practice stage before delivery.

A further application of risk curves is to use them for monitoring an ongoing project. Successive estimates of the dates of occurrence and risk levels of the same milestones provide a very graphic way of tracking how the opinions and confidence of Engineering groups change in the light of later events and new

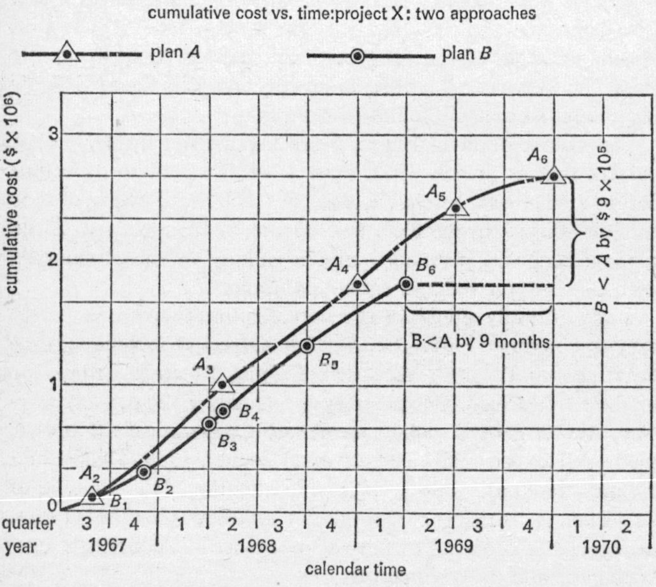

Figure 6

information. Consider the shaded area on the plan B curve on Figure 5. This reflected a revised risk assessment of the B approach, which had by then been in development for about eight months, as of 2/68. A key technical problem had been encountered, the planned first review had slipped from the orginal 9/67 to 2/68, and it was felt that barring solution of the problem, risk would hold at about 20 per cent until 10/68, where (by rights) it should have been down to 3 per cent. The project group felt that they could recover enough of the eight months slippage to deliver only three months late. (Schedule recoverability is an

Table 2

Project X: Risk Reduction

Period	*Approach A* Risk reduction %	$ × 10⁵ spent in period	$/% risk reduction	Weeks in period	Weeks 1% R. R.	*Approach B* Risk reduction %	$ × 10⁵ spent in period	$/% risk reduction	Weeks in period	Weeks per 1% R. R.
Inception to first review	100 − 60 = 40	1	2500	13	0·3	75 − 50 = 25	1	4000	13	0·5
First review to announcement	60 − 40 = 20	9	45,000	35	1·7	50 − 5 = 45	7	20,000	35	0·8
Announcement to delivery	40 − 0 = 40	17	42,500	87	2·2	5 − 0 = 5	10	200,000	48	9·6

interesting problem in its own right. It is described by Norden (1966), but will not be further treated here.)

It has been our experience that whenever the risk level fails to drop as anticipated, a clue or warning signal has been given implicitly, and management review may well be warranted. The pattern is similar to that experienced in per-cent-progress curves discussed by Hertz (1950). Whenever actual progress falls markedly behind plan, the real management problem is to decide whether to allow the project to continue, change its scope, or cut your losses short and stop the job. A collection of tools such as those described herein, when added to the total set of factors under consideration, can aid in the making of more informed decisions.

References

DELKEY, N., and HELMER, O. (1963), 'An experimental application of the DELPHI method to the use of experts', *Management Science*, vol. 9, no. 3.

ECKMAN, D. P. (ed.) (1961), *Systems: Research and Design*, Wiley.

EPSTEIN, B. (1958), 'The exponential distribution and its role in life-testing', *Industrial Quality Control*, vol. 15, no. 6, pp. 2–7.

FREEMAN, R. (1960), 'A generalized network approach to project activity sequencing', *I.R.E. Transactions*, P.G.E.M., vol. EM-7, no. 3, p. 104.

FULKERSON, D. R. (1961), 'A network flow computation for project cost curves', *Management Science*, vol. 7, pp. 167–79.

HELMER, O. (1967), *Systematic Use of Expert Opinions*, Clearing-House for Federal Scientific and Technical Information, no. AD-662320.

HELMER, O., and RESCHER, N. (1959), 'On the epistemology of the inexact sciences', *Management Science*, vol. 6, no. 1.

HERTZ, D. B. (1950), *The Theory and Practice of Industrial Research*, McGraw-Hill, pp. 167–72.

KELLEY, J. E. JR., and WALKER M. E. (1959a), 'Critical-path planning and scheduling', *Proceedings of the Eastern Joint Computer Conference.*

KELLEY, J. E., and WALKER, M. E. (1959b), *Critical-Path Planning and Scheduling: An Introduction*, Mauchly Associates.

LORGE, I., and SOLOMON, H. (1955), 'Two models of group behaviour in the solution of Eureka-type problems', *Psychometrika*, vol. 20, no. 2, pp. 139–49.

MALCOLM, D. G., ROSEBOOM, J. H., CLARK, C. E., and FRAZAR, W. (1959), 'Application of a technique for research and development program evaluation', *Operations Research*, vol. 7, pp. 646–69.

MARPLES, D. L. (1961), 'The ducts and valves problem of the advanced gas-cooled reactor (A.G.R.) of the atomic energy authority', *I.R.E. Transactions*, P.G.E.M., vol. EM-8, no. 2, pp. 55–70.

McGee, A. A., and Markarian, M. D. (1961), 'Optimum allocation of research engineering manpower within a multi-project organizational structure', *I.B.M. F.S.D. Report*, no. 61–907–171, Federal Systems Division.

Norden, P. V. (1960), 'On the anatomy development projects', *I.R.E. Transactions*, P.G.E.M., vol. EM-7, no. 1, p. 41.

Norden, P. V. (1961), 'The study committee for research, development and engineering (S.C.A.R.D.E.): a progress report', *I.R.E. Transactions*, P.G.E.M., vol. EM-8, no. 1, pp. 3–10.

Norden, P. V. (1966), 'Some properties of project schedule recovery limits', in M. C. Yorits *et al.* (eds.), *Research Progress Effectiveness*, Gorden & Breach.

Norden, P. V., and Bakshi, A. V. (1960), 'Internal dynamics of research and development projects', *Management Sciences Models and Techniques*, Pergamon, pp. 187–205.

Pert (1958), *Program Evaluation Research Task, Phase 1*, Summary Report Special Projects Office, Bureau of Ordnance, Dept of the U.S. Navy.

Voress, H. E., Houser, E. A., and Marsh, F. E. (1961), 'Critical path scheduling' (A Preliminary Literature Search), Bulletin TID-3568, U.S. Atomic Energy Commission, lists 255 items.

Part Two **Process Decisions and Design**

Which do you prefer, a dollar for a new product or a dollar for a new process? The question is unrealistic, of course, but the issue it raises is not. New products capture the consumers' imagination. Productive invention is newsworthy, dramatically exemplifying the dynamic potentials of our ever-expanding technological capabilities. Because of its newsworthiness, we must ask ourselves whether we are only dealing with pseudo-events, that is, events created by the word rather than the deed. (A dollar for neither the product nor the process – a dollar for public relations?) Process innovations are less heralded – providing weaker drama. The consumer doesn't care enough about the reality of a new process configuration to make himself vulnerable to pseudo-claims.

Management spends its public relations funds on new product announcements while investing its capital in process improvements for old product lines. As long as the guide-line is cost and not profit the basic propensity will be to find better ways for making old things. New process configurations, based on *methodology* and not technology, are beginning to alter this situation drastically. The new methodologies accent profit and encourage larger systems' views. The articles in this section represent this point of view.

6 J. F. Magee

How to Use Decision Trees in Capital Investment

J. F. Magee, 'How to use decision trees in capital investment', *Harvard Business Review*, vol. 42 (1964), pp. 79–96.

The decision-tree approach was described by Magee (1964, p. 126) as a way of displaying the anatomy of a business investment decision and of showing the interplay among a present decision, chance events, competitors' moves, and possible future decisions and their consequences. We saw the values of decision trees in exploring a variety of problems – new-product introduction, plant modernization, research and development strategy, outlays for new facilities, and others.

In this article I will discuss how to go about using the decision-tree concept – the skills needed, organization of the analysis, and data required. A case example of a business investment question will serve as a vehicle for discussing aspects of the practical use of decision trees. As this example will illustrate, the new approach leads to different and improved answers to investment-analysis questions. It enables management to take more direct account of: (a) the impact of possible future decisions; (b) the impact of uncertainty; (c) the relative values of present and future profits; (d) the comparative advantages of varying levels of expected profit and corporate flexibility.

Plan of Attack

Our case example involves a company situation which, though fictitious, incorporates a number of elements of cases I have actually worked with.

The management of Prism Paints, Inc., must decide what to do with one of its manufacturing plants which, though adequate in size, is technically obsolete and unable to supply the quality of products required in the current market. Since the plant in question is rather small by modern standards, and it is argued that a plant of its size is at an economic disadvantage, there is considerable managerial controversy over the proper course of action – whether to modernize the operation by construction of better facilities at that location or to scrap the existing plant and supply the area involved from the company's facilities elsewhere, e.g. plants in adjacent states.

The possible alternatives involve significantly different capital expenditures and appear to lead to substantial differences in operating economy as well. Underlying the controversy over what to do is a concern for future product demand in the area served by the plant.

In this and other cases the key steps in building and using a decision tree for investment-project analysis are:

1. *Identification of the problem and alternatives.* The information about the investment opportunity may come from a number of sources, including market analysis, operations research and engineering analysis, research and development work, trend studies, competitive intelligence, and executive imagination. Experience indicates that the more intimate the executive is with his analytical support, the more alternatives can be seen. An analysis team combining some of the key disciplines needed for the problem is very useful if the decision to be made is a major one.

2. *Layout of the decision tree.* This is the formulation of the structure of alternatives underlying the investment decision.

3. *Obtaining the data needed.* Analyses of various sorts are usually required to estimate cash flows and probabilities of uncertain events. This is another function of the analysis team, if such a team exists.

4. *Evaluating alternative courses.* Evaluations frequently lead to a restatement of the problem and a reformulation of the decision tree. A good evaluation will test which alternatives appear desirable in light of the standards used. It will show whether apparent conclusions are sensitive to changes in doubtful or controversial estimates. It will examine the effect of choosing alternate standards, which in turn may lead to revision of standards, further analysis, or reformulation.

Let us examine each of these steps in this article. We can then compare the decision-tree approach with other approaches, review various problems and questions that may arise, and see how the Bayesian method of probabilities analysis can be combined with decision trees to analyse the worth of further research or market analysis for a project.

Identifying Alternatives

Experience has demonstrated the importance in investment studies of identifying what alternatives, what freedom of action, and what uncertainties exist now and in the future; of estimating the costs, demand volumes, prices, and competitive action anticipated under alternatives; and of narrowing down estimates of the probabilities or likelihood of uncertain events. All future possibilities cannot be identified, of course, but experience shows a reasonable job can be done. The better the job of identifying and estimating, the more intelligently can the investment decision be made.

Marketing analysis, operations research, engineering analysis, and financial analysis have vital roles in investment analysis. To be most effective, however, they should be integrated; that is, management should use them *in combination* to explore alternatives and to open up new possibilities.

Those engaged in the analysis should be encouraged to express doubts and uncertainties and to express estimates of costs, technical feasibility, or forecasts of market conditions in terms of *ranges* or *probabilities*. Much as management might wish the uncertainties would go away, undue criticism of the analyst for being imprecise or insufficiently firm in his estimates will only force the uncertainties and the risks underground. The purpose of the investment analysis is to help management identify alternatives and bring out the facts about them.

Innumerable examples could be given of how staff analyses have helped in the identification of alternatives in various companies. I shall mention just a few in order to indicate the *range* of possibilities:

1. One company needed a ten-year forecast of the size and character of the market for new labor-saving equipment. The

forecast was made by combining the results obtained – first, by an industry sector analysis to estimate how individual industries might be expected to grow in response to primary consumer demand and secondary industrial investment activity; second, by a productivity analysis to estimate productivity increases and labor substitutions needed if demand estimates were to be achieved; and, third, by technical analysis of industry operations to see where labor substitutions of the greatest magnitude were most likely to occur. And out of this came a picture with both a range and a depth that otherwise would have been lacking.

2. A chemical manufacturer faced the need to expand capacity to meet growing near-term demand. Additions to capacity were extremely expensive, and the manufacturer was uncertain about the longer term load on the plant. Analysis showed that the manufacturer had other alternatives – e.g. revised scheduling procedures combined with a moderate investment in warehouse space would yield an effective increase in production capacity at modest fixed investment, but at some increase in operating cost. Again, the result of the investigation was a broadening in the range of choice for management.

3. In a study of U.S. Mint facilities investment, the analysis team investigated a series of alternative combinations of plant sizes and multiple-shift operations, to show the trade-off between operating cost and investment required; that is, the savings in operational expenses that could be achieved by increases in outlays for new facilities. In addition, the costs of excess capacity, overtime, and occasional extra shifts were analysed to determine the justifiable amount of excess capacity to handle short-term fluctuations in coinage demand.[1]

4. A proposed field order- and sales-recording system being considered by the management of one company was only marginally attractive. The major part of the investment was in field recording equipment, of which only about half was to be in use more than half the time because of peak-load problems. Investment for handling the peak could be cut, but only with higher operating costs and some congestion. A team of analysts

1. Hearing before a subcommittee of the Committee on Banking and Currency, U.S. Senate, S.874, on 26 March 1963, concerning additional Mint facilities.

incorporating several types of skill was able to suggest an equipment redesign which permitted more intensive use of expensive electronic gear at peak periods, cut total investment by 25 per cent, and made the investment attractive. The original design engineers were unfamiliar with the size of the peak load and its financial impact, so they had not tried to design around it. The analysis team, in turn, might have missed the redesign alternative if it had not included men with engineering experience.

Analysts trained in marketing, operations research, engineering, and finance have a real contribution to make. However, these disciplines need to be integrated in the project analysis if management is to get a clear picture of the available alternatives and their consequences. Figure 1 is a schematic portrayal of the right and wrong ways to think of this process.

Too often the analysis and decision process on a new investment opportunity seems to look like this (the detailed form, of course, depending on the specifics of the project):

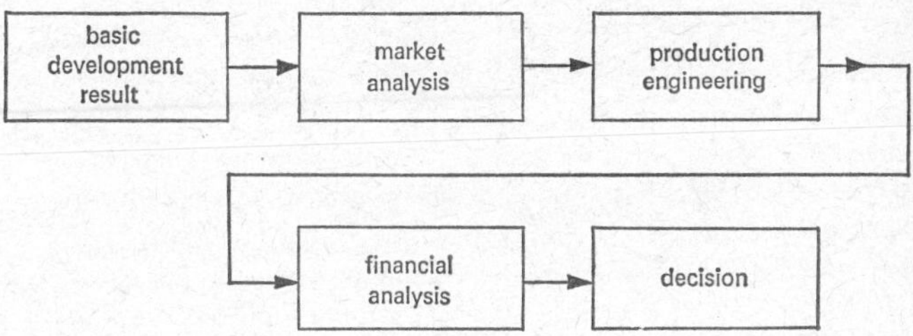

Figure 1 Use of functional specialists in investment analysis

There may be some interaction between market analysis and production engineering or between production engineering and financial analysis, but not a great deal.

The type of interaction management should ask for instead looks more like the diagram Figure 2 overleaf.

Returning to the problem of Prism Paints, let us assume that management assigns the plant-investment question to a group made up of an operations research man, a marketing researcher, and an engineer with practical experience and imagination in process-plant design and costing. This team works closely with the company treasurer to investigate alternatives. It also meets regularly with other members of top management to review progress and policy problems.

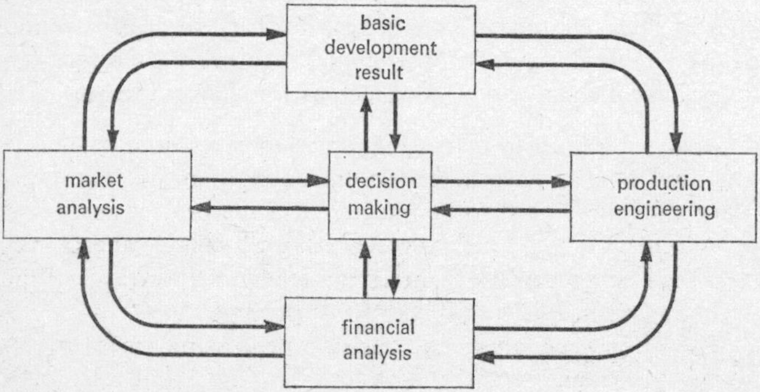

Figure 2

Layout of the Tree

Any significant problem can be examined at several levels of detail. The problem in laying out the decision tree is to strike the right level, the one which permits executives to consider major future alternatives without becoming so concerned with detail and refinement that the critical issues are clouded. The illustrations of trees in this article indicate the general level of detail that seems reasonable in many typical circumstances.

What time span should be covered for the tree as a whole? The layout of alternatives might go on interminably and be carried out over an indefinite future time. Roughly, the practical time span or 'horizon' to consider should extend at least to the point where the distinguishing effect of the initial alternative with the longest life is liquidated, or where, as a practical matter, the differences between the initial alternatives can be assumed to remain fixed.

The time span over which the analysis must extend will vary for *particular decisions.* For some decisions it will be a year or two; for others, ten years or longer. The launching of a new product is an example of the short-horizon type; the net effect of introducing the product, if successful, would last a good deal longer, but it can be characterized as a stream of income continuing for some period at some consistent level, and hence does not need to be portrayed on the tree. There are no visible significant decisions to be considered beyond the first one or two years. On the other hand, plant investment questions, such as Prism Paints' problem of rehabilitating an old plant or building a new one, often have a time span of from five to fifteen years.

Outlining major choices

With these generalizations in mind, let us return to our case. Let us suppose the analysis team begins by getting the forecast of demand shown in Figure 3. The central curve shows the most

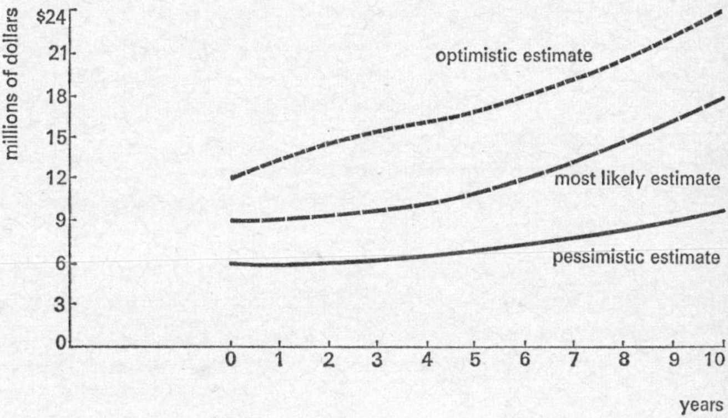

Figure 3 Demand forecasts

likely demand estimate, starting at $9 million in the early years of operation, rising to $12 million in five years, and reaching $18 million at the end of a decade. The lower curve, starting at $6 million and growing to $10 million is the minimum, pessimistic, estimate. The upper curve represents the most optimistic estimate of the rate of market growth.

The team's examination of the demand estimates indicates that there are three basic patterns of operation offering promise:

Program A: To modernize the plant in question and expand elsewhere.

Program B: To close the plant in question and expand elsewhere.

Program C: To modernize and expand the plant in question.

The management, in discussions, points out that one or two major changes in the manufacturing and distribution pattern over the course of the next ten years will be all the company wishes to make, if that many.

As a practical matter, experience also indicates that breaking up a question into two to four decision stages (with the associated uncertainty alternatives) is reasonable. Thinking out the decision at least through a second decision stage enriches the analysis considerably over a conventional single-stage consideration; but, after about four or possibly five stages, the analysis will bog down in complexity and lack of data. This means that the two, three, or four levels or stages of decision points must be the major, significant points of choice as now seen. It means that the time between stages will vary in a given problem and, of course, will vary from one problem to another, depending on the over-all span of time under consideration.

Costs and feasibility

As a next step the analysis team refines the alternatives and makes cost analyses of operations under these systems at different demand levels. Costs are worked out by calculating the best choice of warehouse locations, assignment of warehouses to plants for service, choice of item-stocking policy, transportation method, and so forth, under the three alternate patterns of plant operation and volume of demand. Analysis shows that:

1. Program A is less expensive when demand is less than $10 million annually.
2. Program B is less expensive when demand is between $10 million and $14 million annually.
3. Program C is less expensive when demand is over $14 million annually.

It is decided, therefore, to break up the total time span under consideration into three periods – the first three years, the middle four years, and the period from the eighth year onward – as illustrated in Figure 4. During the first stage, demand is likely to be at a level where Program A would be preferred; during Stage 2, demand is likely to be in a range where Program B would be preferred; thereafter, demand is likely to be high enough so that Program C would be preferred.

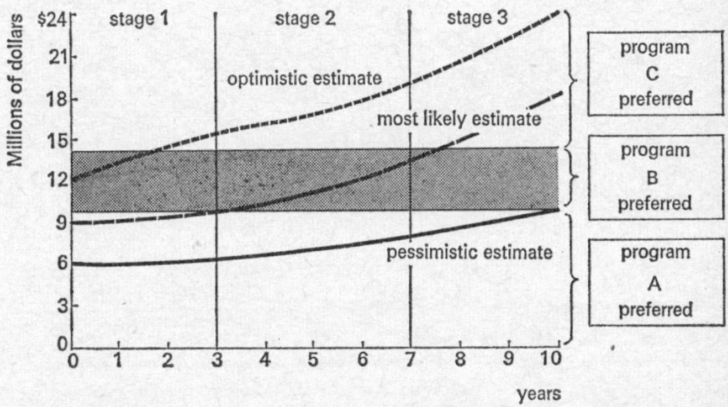

Figure 4 Effect of demand on preference for alternative patterns

In theory there are twenty-seven plans over time that might be considered, although in practice some are out of the question. For example, Prism Paints' management may conclude that if Program B were adopted, it would be impractical to go to A or C, and to re-open the plant in question.

The possibilities can be reduced for practical purposes to thirteen feasible program sequences, each with its own pattern of investment and operating cost. For example, it is possible to begin to follow Program A, and at a later time – say, eight years later – convert to Program C at some cost of sunk investment, rearrangement, and excess operating expense. The thirteen feasible sequences are shown in Figure 5.

Preliminary Drawing

If the company were sure of the sales volumes and operating costs five and ten years hence, it could choose the particular program for plant construction that would make profits largest,

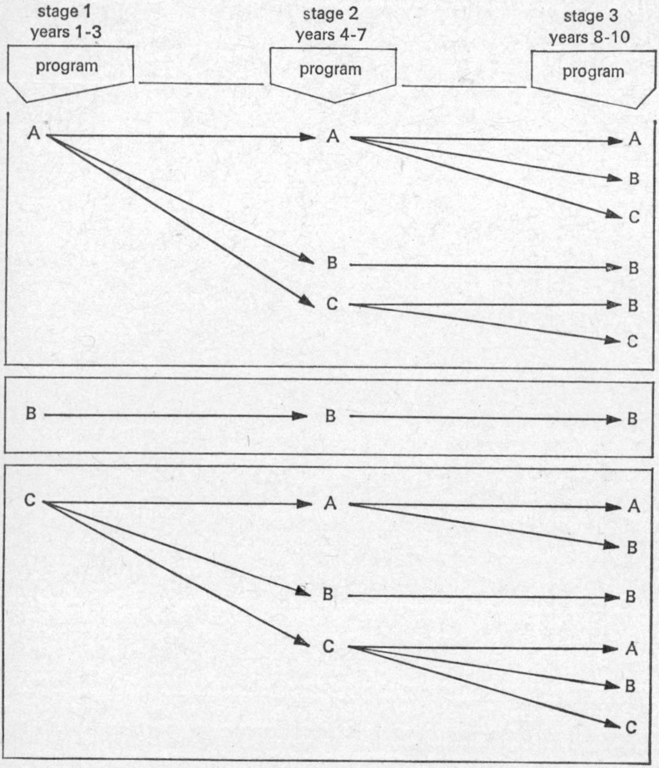

Figure 5 Program sequences

after giving weight in some fashion (for example, by discounting future cash flows) to the relative value of present and future profits. Perhaps if sales do develop as projected, Program C is the least expensive over the next decade. But what if sales do not develop as well as, or grow even faster than, forecast? The task of the team is to allow for the relative flexibility of various programs as a factor in meeting uncertain product demand. The choice made now will decide what later alternatives are still feasible. For example, as mentioned earlier, if Program B is chosen, the company will have to live with it for some time.

What immediate choice is the best one in view of the interrelationships among the immediate choice, demand levels, and future alternatives?

Figure 6 gives a preliminary sketch of the decision tree which shows the interaction between Programs A, B, and C concerning plant facilities and the levels of demand in the company's market area. Some of the alternatives are not carried out to complete detail but are indicated by the dashed lines. The portion of the tree shown represents ten of the thirteen alternative patterns, many of which would be repeated in the portions of the tree that are not shown.

Obtaining Data Needed

In my view the appropriate standard for corporate investment decisions is the maximization of expected wealth, or of discounted expected cash profit where future cash profit is discounted to present value at a discount rate equivalent to the market 'cost of capital' of enterprises of similarly uncertain future (Magee, 1964, p. 133). I recognize, of course, that not all businessmen agree with this standard. Those who do not agree can use the principles to be described, but they will want to adjust them to suit their own criteria. Just bear in mind that in *this* illustration which I am presenting the yield figures to be shown on the decision tree are not accounting profits but cash flows.

The numbers or values needed at each stage under the standard of maximum expected cash profit are:

1. The probabilities of each of the alternative uncertain outcomes – in Figure 6, for instance, the probabilities that demand will be low, medium, or high during years one through three.
2. The cash flow associated with each combination of decision alternative and chance outcome.
3. Where the time span of the stage is significant – of the order of a few months or more – an estimate of the discount rate to be applied to deferred cash flows in the stage or to the future 'position value' of the outcome. (See the discussion of rollback in Magee, 1964, p. 132.)

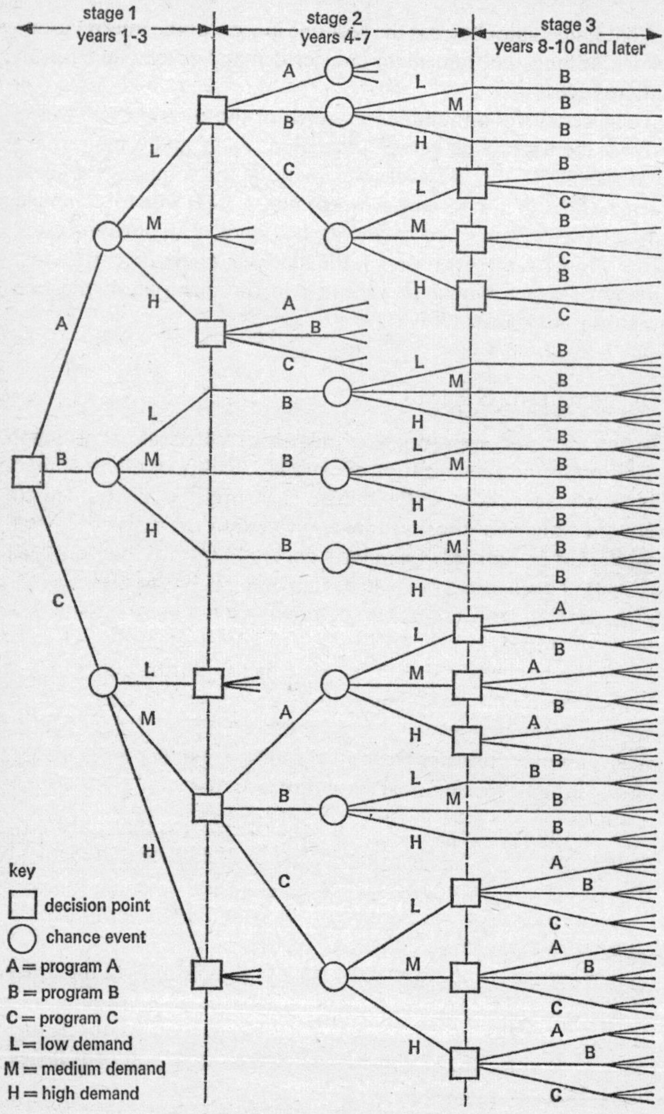

Figure 6 Preliminary decision tree for Prism Paints, Inc.

Probability estimates

The probabilities of the uncertain alternatives may sometimes be estimated objectively from research. For example, the likelihood of surges or slumps in demand may be estimated objectively from a statistical analysis of demand variations, or the possibility that the market for a proposed product will be one size or another may be estimated from survey data. At other times the probabilities may have to be estimated subjectively from the intuition of an experienced operator or staff man. For example, the likelihood that a development project will succeed may be estimated from the intuitive judgments of two or three skilled engineers or research managers.

What is gained by trying to make subjective estimates of probabilities where limited objective information exists? Why not make the decision on feel, hunch, or intuition in the first place? For one thing, the estimation of elemental probabilities (and other values) permits various parties to the decision to see the basis for each other's conclusions. For another, it permits the executive to make use of the intuitions and skills of subordinate operators as staff members without abdicating his position as the decision-maker. Further, it permits analysis of the impact of variations in the estimate – what is called *sensitivity analysis* – in the conclusion. Finally, as noted later, it provides a means for measuring the value of further work to sharpen up the estimates. This step is called *Bayesian analysis.*

Let us assume that the need for these estimates is felt by the managers in our Prism Paints case. Accordingly, the team analyses evidence on market demand and the data underlying the demand forecasts (see Figure 3) to estimate how likely it is that demand will fall in the low, medium, or high range in each of the three stages of the analysis. The team concludes that these estimates must be made in relation to demand in the preceding stage. That is, if demand is high in Stage 1, that will certainly influence the likelihood that demand will be high in Stage 2. These estimates, for each stage, are summarized in Table 1.

Table 1

Probabilities that Demand will be Low, Medium, or High

		Stage 2			*Stage 3*		
		If Stage 1 is:			*If Stage 2 is:*		
Level of demand	*Stage 1*	*Low*	*Medium*	*High*	*Low*	*Medium*	*High*
Low (under $10 million)	0·50	0·35	0·15	—	0·20	0·05	—
Medium ($10–$14 million)	0·43	0·50	0·45	0·40	0·60	0·35	0·20
High (over $14 million)	0·07	0·15	0·40	0·60	0·20	0·60	0·80
Total	*1·00*	*1·00*	*1·00*	*1·00*	*1·00*	*1·00*	*1·00*

Cash flows

The cash flows are characteristically estimated from the types of marketing, operations, engineering, and financial analyses mentioned earlier. Where the stage covers a considerable time span – one or more years – the schedule should show the cash flow at least on an annual basis.

In our Prism Paints case, distribution analyses show that the relative operating profitability of the alternatives facing management can be expressed by net annual cash flow as follows (in millions of dollars):

	Demand level		
	Low	*Medium*	*High*
Program A	$1·5	$1·8	$1·5
Program B	1·0	2·0	2·8
Program C	0·5	1·5	3·3

Initial investment to implement Program A is estimated to be $12 million, compared with $14·75 million for Program B or $14 million for Program C. Program A could thereafter be converted to B at a cost of $6 million, or to C for $8 million. Program C could be converted to A for $4 million, or to B for $6 million.

The management of Prism Paints establishes a desired return

on investment or cost of capital of 14 per cent per year. This is the rate to be used to reduce future cash flows to a present value for comparative purposes.

Evaluating Alternatives

We can now construct the complete decision tree for Prism Paints' plant-investment problem. The tree is shown in Figure 7 (ignore for a moment the white numbers outside the boxes and circles).

Part A shows how the three alternatives at the initial decision, when combined with the possible market conditions in the period one to three years ahead, can lead to six decision situations at the beginning of the second stage in addition to the close-down situation (Program B). The remaining parts (B through D) show how three of these decision situations may develop over the course of the subsequent years. (The trees for the four situations not illustrated in Figure 7 would be quite similar in design.)

To evaluate the data in the decision tree, we follow three steps:

1. Evaluate each of the alternatives at the final-stage decision points. Select the alternative with the largest net present value. Assign this value to the position. For instance, in Stage 3 of Decision no. 1 (see Figure 7b), the expected value of each alternative for the uppermost decision can be calculated as shown in Table 2. The alternative with the largest net present value ($12·8 million) is Program A. Therefore, Program A would be chosen *if* this decision at Stage 3 *were* to be made *now*.

2. Evaluate each decision alternative at the next preceding stage. This is accomplished by calculating the expected discounted present value of the cash flow during the stage, including the value of the following stage as if it were a lump sum received at the end of the period. For instance:

At Stage 2, Decision no. 1 (Figure 7b), there are three chance alternatives related to market conditions. When the present value of the cash flow of each alternative – including the following stage values – is calculated and weighted by the likelihood of its occurrence, the expected value of the choice is found. The value of Program A is found to be $12·6 million. Similarly, the value

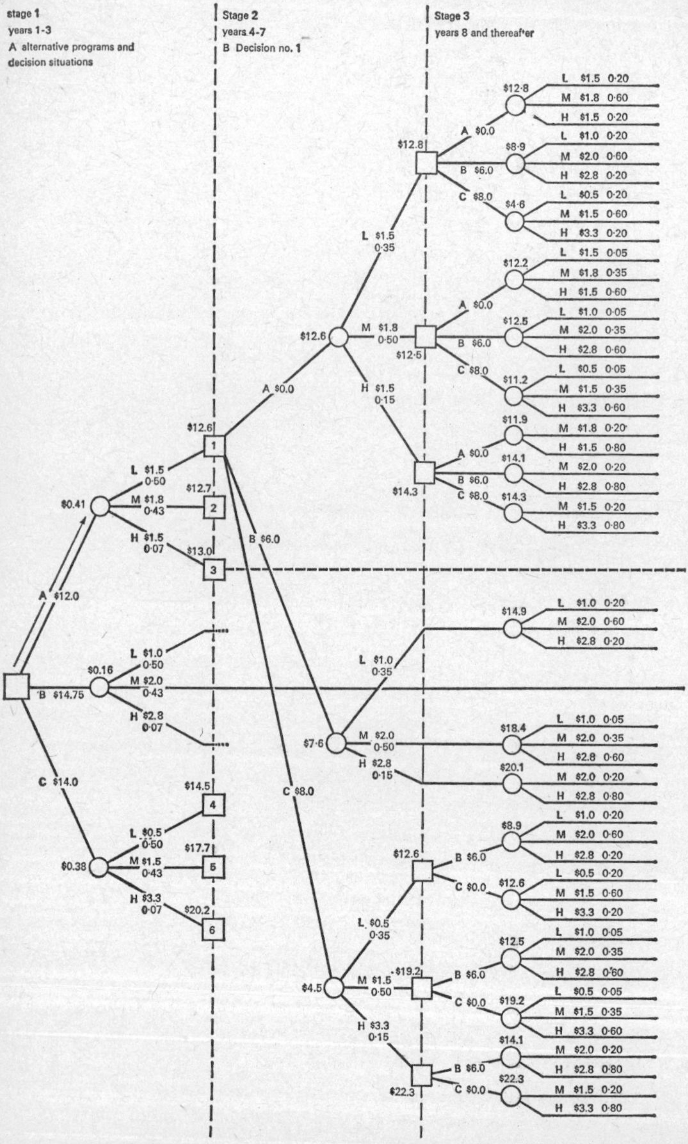

Figure 7 Complete decision trees for Prism Paints, Inc.

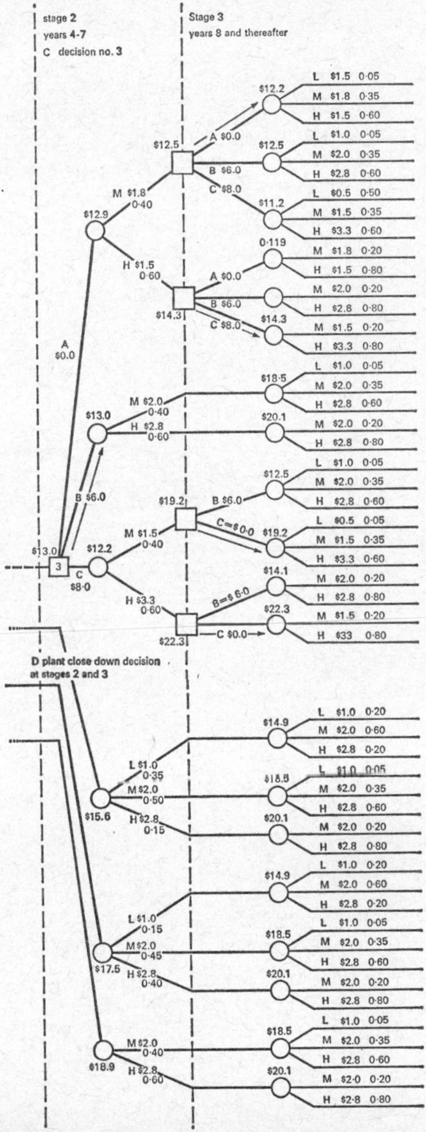

note:

The choices at each decision point are indicated by the letters A, B, or C.

The figures following these choices represent the investment required.

The possible levels of demand (low, medium, and high) are indicated at each chance-event point by the letters L, M, and H.

Two sets of numbers go with each market level – a dollar figure representing cash flow and a decimal representing the estimated likelihood that the level will come about.

The numbers outside the decision-point boxes and the chance-event circles represent present value (in millions of dollars).

The arrow at a decision box points to the alternative with the greatest present value.

key

□ decision point

○ chance event

A = program A

B = program B

C = program C

L = low demand

M = medium demand

H = high demand

Table 2

Present Values of Alternatives for a Decision at Stage 3 (Dollar figures in millions)

Alternative	*Market size*	*Likelihood* (1)	*Cash flow per year* (2)	*Period*	*Present-value factor** (3)	*Total present value* (1) × (2) × (3)
Program A	L	0·20	$1·5	Indefinite	7·6	$ 2·3
	M	0·60	1·8	Indefinite	7·6	8·2
	H	0·20	1·5	Indefinite	7·6	2·3
					Total	*$12·8*
					Less investment	0·0
					Net present value	$12·8
Program B	L	0·20	$1·0	Indefinite	7·6	$ 1·5
	M	0·60	2·0	Indefinite	7·6	9·1
	H	0·20	2·8	Indefinite	7·6	4·3
					Total	*$14·9*
					Less investment	6·0
					Net present value	$ 8·9
Program C	L	0·20	$0·5	Indefinite	7·6	$ 0·8
	M	0·60	1·5	Indefinite	7·6	6·8
	H	0·20	3·3	Indefinite	7·6	5·0
					Total	*$12·6*
					Less investment	8·0
					Net present value	$ 4·6

* The value of $1 per year, received for an indefinite period and discounted at 14 per cent per year on a semiannual basis, is $7·60. Thus, 7·6 times the annual cash flow is the present value of the cash flow discounted at 14 per cent per year.

of B is found to be $7·6 million, and the value of C is found to be $4·5 million. (The mathematics of this step are the same as described in my July–August article.) On this basis, the preferred choice at Decision no. 1, Stage 2, is therefore Program A, and the position value of Decision no. 1 is $12·6 million (Magee, 1964, p. 132).

3. By repeated application of this process – the process of 'rollback' (referred to in my previous article (1964)) – to each stage of decisions, the value of each alternative at the first stage can be found. In Figure 7a we find these values for the three programs at Stage 1: A, $410,000; B, $160,000; and C, $380,000.

These are the amounts by which the expected present-value cash flow exceeds or falls short of the initial investment called for under each alternative. The cash flows under Programs A and B are more than sufficient to earn the required 14 per cent on the investment; the cash flow from Program C is insufficient.

Program A is the preferred choice because it has the greatest value. This does not mean, however, that Prism Paints must live with Program A indefinitely. If the market grows rapidly to high levels, management can shift to Program B (closing the plant, expanding elsewhere) at the intermediate point. Otherwise management can choose depending on the rate of market development during the next seven years or so.

Actually, of course, during the years ahead management will learn more about the market, and conceivably other changes will take place. Thus the *immediate* choice among A, B, and C is all that must be made now. However, the immediate decision is made with a proper eye to its impact on future developments and possible later decisions – with a recognition of the 'futurity' of the current decision.

Comparing Other Methods

Does the use of the decision-tree approach make any difference? Let us compare the result of the Prism Paints analysis with the results of other more conventional discounted cash-flow techniques.

One common method is to evaluate each fixed alternative

under the 'most likely' cash flow using present-value techniques. This gives us (in millions of dollars):

		Cash flow per year		
	Investment	*Years 1–3*	*Years 4–7*	*Year 8 and thereafter*
Program A	\$12·00	\$1·5	\$1·8	\$1·5
Program B	14·75	1·0	2·0	2·8
Program C	14·00	0·5	1·5	3·3

This approach yields net present values as follows (when the cash flows are discounted at 14 per cent per year): Program A, 0; Program B, \$1·3 million; and Program C, \$2·1 million.

It would appear that Program C is the preferred; this ignores the possibility that the market may not in fact develop as fully as the 'most likely' levels.

Another approach is to calculate the net *expected* present value, with the cash flows weighted by the estimated likelihood of their occurrence. This technique, applied to the Prism Paints data, yields the following net present values: Program A, \$126,000; Program B, \$160,000; and Program C, \$380,000.

This approach seems to show Program B as somewhat superior to A. This is due to the fact that the technique does not take into account the opportunity, if A is chosen, to modify the distribution plan later. As the decision-tree analysis shows, the flexibility implicit in Program A makes it significantly superior.

The decision-tree analysis brings out the impact both of uncertainty and of possible future decisions, conditioned on future developments. To emphasize a point made earlier, the analysis *could* be made without actually drawing a tree, i.e. by simply listing the alternatives and stages for computation instead; but in practice the visual form has real value in helping managers understand and discuss an investment problem.

Special Problems

In the Prism Paints case example, I have used 'discrete' chance events for illustrative convenience. A discrete chance event is one where there is some number of clear-cut chance outcomes, each with its probability of happening. For instance, the toss of a coin is a discrete chance event.

In many practical cases, however, the chance events are in fact continuous; a whole range of answers is possible. There may be an estimated probability distribution associated with this range. The daily temperature is an example of a continuous chance event. There is an expected value for any locale and any time of year, though the actual value can be anywhere in a wide range. Using a probability distribution related to temperature, you can speak of the chance the temperature will be between 75° and 80°, below 65°, and so forth.

Continuous-chance variables can be handled readily using the decision-tree technique. There are two ways:

1. In the Prism Paints case, we arbitrarily sliced the continuous-chance variable (future market levels) into the form of discrete chance levels – low, medium, and high. This approach often is adequate. It oversimplifies, but (in many cases) not so much so as to add serious error.

2. We can calculate the values of the decision alternatives at any stage as related to the full range of the prior chance outcomes. In this way we can fix the part of the range of possible outcomes for which any alternative is best. In some cases, straight calculation is all that is needed; in other cases, simulation methods are useful (for example, where the relationship between the chance variable and the cash flow following some decision alternative is a complex one).

Another problem is that of many chance variables. In the Prism Paints example, I have used market level as the single-chance variable at each stage, but in some problems there are several uncertain variables at each stage. These might include, for example, total demand, share of market, costs, yield, or results of a development or promotion program. Any one or all of them might be so uncertain that a range of probabilities should be used.

Uncertainty connected with several variables can be handled readily. One useful way is the 'Monte Carlo' method, a form of simulation:

At each stage – still working back from the last stage – values of the chance variables are picked at random, and the value of each decision alternative is worked out. Through repeated application, the expected values and probability distributions for each decision alternative emerge. The best alternative is chosen, and the 'rollback' principle is applied to carry out the 'Monte Carlo' analysis, if needed, at the next preceding stage.

Finally, there is the calculation problem. The limiting factor in drawing up a complex decision-tree analysis is not the task of computation but the capacity of the analysts to imagine alternatives and think out the implications of the various possible choices. Even so, computer calculation is often useful, and it has one important advantage: it makes it easier to analyse the sensitivity of the end conclusion to changes in key investment or cash-flow figures or in probability estimates. If the conclusion as to what to do now is not too sensitive to changes in certain estimates, these can be estimated roughly. If the conclusion is very sensitive to changes in certain numbers, these are the ones where further research may be in order.

Use of Bayesian Method

One alternative that frequently exists but may not be considered is further research or investigation. This does not mean procrastination but an intensive look by a fresh and objective team of men. Such a look may be especially valuable where the existing basis for decision is a series of semi-independent studies by staff groups committed to a position. It may mean carrying a research or development program one step further, or making a marketing study to try to narrow the range of market uncertainty. The concepts of Bayesian statistics (Schlaifer, 1959) give a means for building in subsequent information to modify estimates of probabilities. The Bayesian method also gives a means for estimating the value of further investigation.

In the previous article in this series, we discussed the investment issue facing Stygian Chemical Industries, Ltd – whether to build a small plant or a large one, in view of the uncertainty concerning the size of the market. The key issue facing management was whether to gamble on the size of the market and build a large plant immediately or to build a small plant with the possibility of expanding it later. That example lends itself particularly well to the present question, so I shall go back to it now.

Perfect foresight?

Using discounted expected cash profit as the criterion and discounting future cash flows at a 10 per cent rate per year, we found in the previous article (Magee, 1964) that the appropriate decision for Stygian appeared to be the large-plant alternative. The discounted expected value of that alternative was $1·47 million, compared with a $1·18 million value for building the small plant (Magee, 1964, pp. 133-4).

Suppose the company considers now the possibility of doing some further research – for example, making an independent technical–economic or marketing study to estimate more carefully what the long-term market penetration possibilities are. The marketing study properly and thoroughly done will be expensive; it will cost $100,000. Should the company contract for the research?

First, what would perfect foresight be worth? If management *knew* the market would be large, it would build the large plant. If it *knew* the market would be small in the long run, it would build the small plant. The value of knowing for sure whether the market is large or not can be calculated as follows:

1. Assuming one of the anticipated market conditions occurs and the preferred strategy for it is followed, calculate the present value of the cash flow which will come to the company (subtracting the cost of plant investment).
2. The discount rate used depends on management's judgment of the situation. For our illustration here, let us assume a 10 per cent annual discount rate on future cash flows.
3. Multiply the present value thus obtained by management's present estimate of the probability that the market condition will actually occur.
4. Repeat the procedure for the other market conditions anticipated.

These calculations are carried out in Table 3. They give us, in effect, a kind of average of the discounted cash flows weighted by the likelihood of their occurrence, if management knew in advance how big the market would be. This result ($2·79 million) is greater than the discounted expected value of the preferred large-plant alternative ($1·47 million) by $1·32 million. Thus, the value of perfect foresight compared with present uncertainty is $1·32 million.

Table 3

Calculations for Determining Value of Perfect Foreknowledge (dollar figures in millions)

Market situation (demand)	*Preferred strategy*	*Present value* of preferred strategy (1)*	*Present estimate of likelihood (2)*	*Value of perfect foresight (1) × (2)*
High average	Big plant	$3·7	0·60	$2·22
High initial, low average	Small plant	1·5	0·10	0·15
Low average	Small plant	1·4	0·30	0·42
Total				$2·79

* The present value of the cash flow, discounted at 10 per cent annually, including plant investment, and assuming the specified market condition occurred and the preferred strategy was followed.

New alternatives

This estimate suggests that there is a sizable potential pay-off from good research in reducing the likelihood of a mistake, by narrowing down uncertainty in the market estimates, and thus increasing the expected profit from the action taken. Let us suppose management has confidence in the research team, though considering it conservation. Management estimates that:

1. If the market actually turns out to be large and demand is high from the outset, the chance the research will show a positive result is 70 per cent. (By 'positive result' I mean a high estimated demand.)

2. If demand is high initially but tapers off to a low level, the research effort might be misled, and there is a 50 per cent chance it will err and report a positive finding.

3. If, however, demand is low, the chance of a positive result is quite small – about 5 per cent.

From these estimates it is possible to calculate that:

1. If the research result is positive, the chance of: (a) a high average demand is 0·87; (b) an initially high, low subsequent demand is 0·10; (*c*) a low average demand is 0·03.

2. If the research result is negative, the chance of: (a) a high average demand is 0·35; (b) an initially high, low subsequent demand is 0·10; (c) a low average demand is 0·55.

3. The advance estimate of a positive finding is 0·485. (The difference between the estimate of 0·485 that the research will show a large demand and management's estimate of 0·60 that the demand will be large is a measure of management's estimate of the research organization's degree of conservatism.)

For those interested in technique, the method of calculating the foregoing data is shown in Table 4.

Table 4

Method of Obtaining Estimates of Research Reliability

Let HH signify a high average demand; HL signify an initially high, but low subsequent demand; and LL signify a low average demand.

Let F represent a positive result from the research and N a negative result.

Then:

$$P(HH/F) = \frac{P(F/HH)P(HH)}{P(F/HH)P(HH) + P(F/HL)P(HL) + P(F/LL)P(LL)}$$

$$= \frac{0{\cdot}7 \times 0{\cdot}6}{0{\cdot}7 \times 0{\cdot}6 + 0{\cdot}5 \times 0{\cdot}1 + 0{\cdot}05 \times 0{\cdot}30} = 0{\cdot}87$$

The probabilities P(HL/F), P(LL/F), P(HH/N), P(HL/N), P(LL/N) can be obtained similarly.

The chance of the research showing a positive result is:

$$P(F) = P(F/HH)P(HH) + P(F/HL)P(HL) + P(F/LL)P(LL) = 0{\cdot}485$$

Now management has three alternatives: (1) to build a large plant; (2) to build a small plant, with the later possible option of expansion at Decision no. 2; or (3) to commission the research. If the third alternative is chosen, then, depending on the outcome of the investigation, the decision alternatives are 1′ and 2′ at

Decision no. 3 (see Figure 8) if the research outcome is favorable; or 1″ and 2″ at Decision no. 4 if the research outcome is unfavorable. These correspond to the original alternatives: (a) building a large plant; or (b) building a small plant.

While the modified decision tree looks a good deal more complex, careful examination will show that the portion following Decision no. 3 and the portion following Decision no. 4 are each duplicates of the original decision tree, with just the probability estimates changed. The probability estimates are changed, as earlier described, to reflect the results of the research and its estimated reliability.

Maximizing expected value

The calculation of position values associated with the various decision points can be carried out just as shown in my previous (1964) article. Using discounted expected value as the criterion and a discount rate of 10 per cent annually, one finds that:

1. If management were at Decision no. 3 (research carried out, positive finding), the indicated decision would be to build the big plant, and the discounted expected value – the position value – would be \$3·1 million.

2. If management were at Decision no. 4 (research carried out, negative finding), the indicated decision would be to build the small plant and hold to it, whether initial demand were high or low. The discounted expected value, or position value, would be \$2·6 million.

The expected value of Alternative 3 at Decision no. 1, therefore, can be calculated as:

0·49 × \$3·1 million =	\$1·5 million
Plus 0·51 × \$2·6 million =	\$1·3 million
	\$2·8 million
Less cost of research	0·1 million
Net expected value, Alternative 3	\$2·7 million

The discounted expected profits from Alternatives 1 and 2 are unaffected by the introduction of the third alternative. So Alternative 3, with a value of \$2·7 million, is clearly superior to the previously indicated choice, Alternative 1, with an expected value of \$1·5 million.

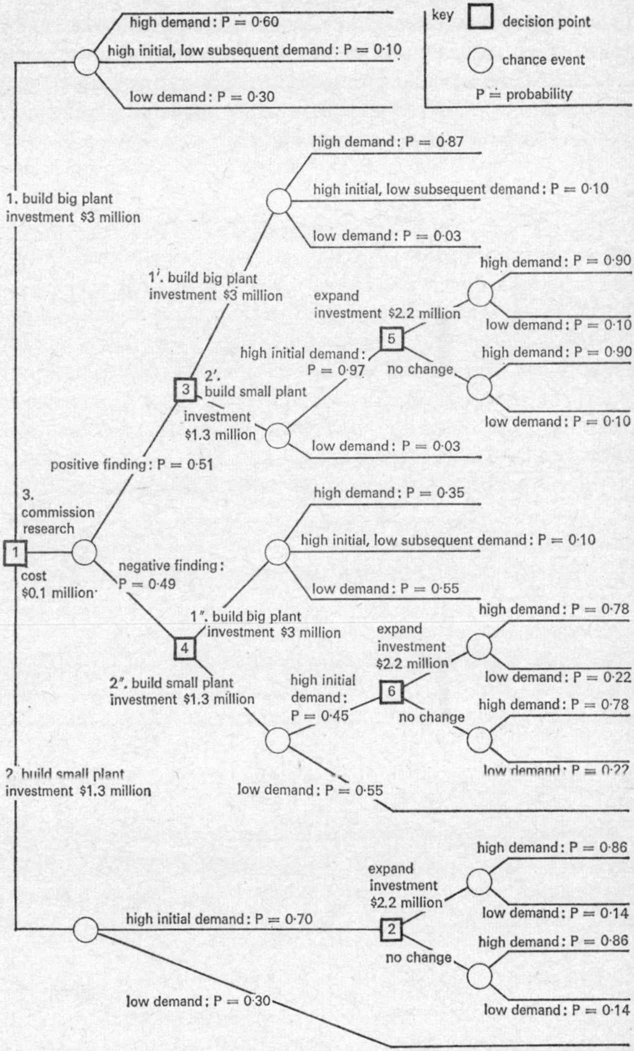

Figure 8 Decision tree for Stygian Chemical with research alternative

The research increases the expected value of the endeavor by $1·2 million. It accomplishes this by reducing the uncertainty surrounding the demand estimates. It is interesting to note that the expected value of the endeavor will be increased significantly *no matter how the research comes out.*

Conclusion

Use of the decision tree concept as a basis for investment analysis, evaluation, and decision is a means for making explicit the process which must be at least intuitively present in good investment decision making. It allows for, indeed encourages, revision of the issue and maximum use of analysis, experience, and judgement. It helps force out into the open those differences in assumptions or standards of value that underlie differences in judgement or choice. It keeps the executive from being trapped in the formalisms of a rigid capital evaluation procedure in which there is little room for feedback, redefinition, or interplay between analysis and decision

The decision-tree approach will strike some businessmen as complex. Certainly it is more complicated than rule-of-thumb approaches, but of course any realistic method would be. In reality, a decision tree need be only as complex as the decision itself. If the decision is a simple choice among alternatives, then the decision tree reduces to a single-stage analysis, i.e. the use of the present-value technique applied to alternative cash flows. If the situation is more complicated, more stages and alternatives are necessary to reflect that fact.

Explicit use of the decision tree concept will help force a consideration of alternatives, define problems for investigation, and clarify for the executive the nature of the risks he faces and the estimates he must make. The concept can thereby contribute to the quality of decisions which executives, and only executives, must make.

References

MAGEE, J. F. (1964), 'Decision trees for decision making', *Harvard Business Rev.*, vol. 42.

SCHLAIFER, R. O. (1959), *Probability and Statistics for Business Decisions*, McGraw-Hill.

7 H. M. Weingartner

Equipment Replacement Analysis: A Note on the Optimum Investment Period

H. M. Weingartner, 'Equipment replacement analysis: a note on the optimum investment period', *Industrial Management Review*, vol. 7 (1965), pp. 47–50.

In the analysis of replacement decisions, one of the quantities to be determined is the optimal life of all challengers[1] so that the cost of the replacement policy involving a challenger may be obtained and the least costly alternative chosen. A replacement policy, rather than a specific replacement decision, is needed since the given capacity for production may be required either for a specified period of time or for the indefinite future. It is the purpose of this note to relate the analysis of the optimal replacement life in the latter case to a problem of capital theory, the optimal investment period, and to illustrate the logic of the underlying calculations in a new way.

A problem familiar to all economists is that of finding the optimum length of time to allow each tree in an infinite series of trees to grow so as to maximize the net income stream generated thereby. Given an investment outlay of C (for purchasing the sapling and planting it), it is assumed that an entrepreneur can realize proceeds of $V(t_1)$ by cutting down and selling the tree at time t_1 without incurring any expenses in the intervening period. Furthermore, it is assumed that he can invest an amount C again at that time, and after an additional t_2 years he can realize proceeds of $V(t_2)$. and so on.

The solution to the problem is usually obtained by solving an equivalent problem, that of maximizing the present value of the chain of investments, although usually it is also assumed that the life of each tree in the sequence is the same (Lutz and Lutz, 1951, pp. 32–3). Both the constancy of investment life and the present value of the chain can be derived by use of a dynamic programming formulation.

1. i.e. potential replacements to a machine in use, in the language of G. Terborgh (1949).

The Principle of Optimality

Let $\phi(t_1)$ be the maximal present value attainable from the infinite chain of investments if the first tree is cut at age t_1 and an optimal policy is followed thereafter. Using continuous compounding at an assumed constant rate of interest, r, we have

$$\phi(t_1) = V(t_1)e^{-rt_1} + e^{-rt_1}\,\phi(t_2) \qquad \mathbf{1}$$

by the principle of optimality;[2] we see at once that $t_1 = t_2$, i.e. that each tree in succession is cut at the same age. Substituting t for t_1 and t_2 in **1** and transposing we obtain

$$\phi(t) = \frac{V(t)e^{-rt}}{1 - e^{-rt}}, \qquad \mathbf{2}$$

and the optimal t, denoted by t^*, satisfies the first-order condition

$$\frac{dV(t^*)}{dt^*} = \frac{r[V(t^*) - C]}{1 - e^{-rt^*}}. \qquad \mathbf{3}$$

An alternative derivation may make an intuitive grasp of relation **3** somewhat easier. As stated at the outset, the problem may be viewed as maximizing the net income stream derived from the chain of investments. Thus we may suppose that a tree cut down at time t and yielding $V(t)$ permits the owner to consume the present value of the proceeds over the interval from 0 to t after setting aside another amount C for the next tree. The present value of these net proceeds is then

$$[V(t) - C]e^{-rt}, \qquad \mathbf{4}$$

and it yields a constant and continuous stream $R(t)$ for the t years of

$$R(t) = \frac{r[V(t) - C]e^{-rt}}{1 - e^{-rt}}. \qquad \mathbf{5}$$

The term $\frac{re^{-rt}}{1 - e^{-rt}}$ may be recognized as the 'capital recovery factor' or 'the annuity factor', using continuous compounding. Maximization of $R(t)$ then also yields **3**.

Interpretation of the optimality criterion, equation **3**, may be

2. 'An optimal policy has the property that whatever the initial state and initial decision are, the remaining decisions must constitute an optimal policy with regard to the state resulting from the first decision.' R. Bellman (1957, p. 83).

facilitated if we hypothesize another asset, a 'tree' whose cost is also C and whose growth function $G(t)$ is such that *it does not matter when it is cut down and sold and another is planted.* This hypothesized tree will yield the same continuous net income stream independent of its life t (except possibly at $t = 0$);

i.e. $$R(t) = \frac{r[G(t) - C]e^{-rt}}{1 - e^{-rt}} = k. \qquad \textbf{6}$$

The growth in value of the proceeds of this hypothetical tree, $G(t)$, may be exhibited by solving **6**, which yields

$$G(t) = \frac{k}{r}(e^{rt} - 1) + C. \qquad \textbf{7}$$

The important parameter of this constant-income growth function is k, the rate of income generated, and the function may perhaps be denoted better by $G(k, t)$. This suggests the graphical

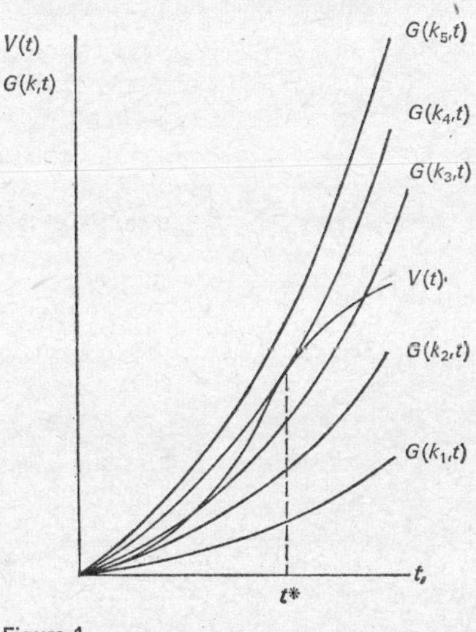

Figure 1

portrayal in Figure 1, in which the function $V(t)$ is assumed to grow first at an increasing rate, than at a decreasing rate, with the resulting family of curves $G(k, t)$ shown. For given k, say

$k = k^0$, the slope of $G(k^0, t)$ is $k^0 e^{rt}$, which increases continuously. An investor is indifferent between points on the function $G(k^0, t)$, but he prefers to be on the highest curve which he can reach. Thus the point of tangency between $V(t)$ and some $G(k, t)$ determines the optimum t.[3] Should there be multiple points of tangency, the G-function with the largest value of k yields the optimum.

Equipment Replacement Analysis

Equipment replacement problems may be analysed by much the same reasoning. In the no-obsolescence case,[4] a machine is assumed to be replaced by an indentical one, *ad infinitum*. Assuming also that capacity is to be maintained,[5] the relevant consideration then is the minimization of the discounted value of all costs required to maintain that capacity. The function $V(t)$ here may be interpreted as the cumulated operating costs. More precisely,

$$V(t) = \int_0^t v(\tau)e^{r(t-\tau)}\, d\tau, \qquad 8$$

where $v(\tau)$ is the rate of operating and maintenance costs, not assumed to be constant. If C is the outlay required to obtain the machine, assumed to remain the same in the future, and salvage value is assumed to be zero,[6] then $\phi(t)$ in **1** is the present value of the policy of replacing this machine with an identical one every t years, and $R(t)$ is the rate of annual outlay required to maintain this policy. Aside from providing the optimal life, t^*, of the challenger, the quantity $R(t^*)$ (often called the equivalent annual investment) may be used to compare different challengers with different optimal lives (Morris, 1960, ch. 9).

3. We are assuming that $V(t)$ is continuous and continuously differentiable.

4. This condition is more restrictive than necessary.

5. I.e. that demand conditions are not altered so as to take the problem out of the realm of pure replacement decisions.

6. Alternatively, C may be taken to be the outlay for replacement after allowance for salvage, also assumed to remain the same. In this instance it would also be necessary that the salvage value of the incumbent machine is the same as the salvage value of the challenger.

In this cost minimization problem, the family of curves $G(k, t)$ in **6** is concave rather than convex as in the diagram, and the cost function $V(t)$ first falls and then rises. The optimal life t^* is then the point of tangency between $V(t)$ and the lowest member of the family, $G(k, t)$.

Role of the Interest Rate

One final point must be brought out – the role of the rate of interest in this calculation. Returning to the simpler tree example, we recall the equivalence between the maximization of the present value of the chain of investments and of the rate of income generated by this chain. The function $G(k, t)$ demonstrated that the equivalence depended on the ability of the investor to convert the periodic proceeds, received every t^* years, into a rate of income which remains constant and is as large as possible. Thus the conversion of a stock into an equivalent flow made use of an interest rate – a rate determined outside the investment itself, presumably by a capital market.

In a long series of publications from a 1935 article (Boulding, 1935, pp. 475–94) to the most recent revision of his well known textbook (Boulding, 1955, p. 870), Professor Boulding has maintained that the objective of the entrepreneur is to maximize the internal rate of return on the investment. This quantity, denoted by ρ, must satisfy the relation

$$C = V(t)e^{-\rho t} \qquad \textbf{9}$$

or

$$\rho = (1/t) \log_e \left[\frac{V(t)}{C}\right]. \qquad \textbf{10}$$

Maximization of ρ in **9** or **10** yields the criterion:

$$\frac{V'(t^0)}{V(t^0)} = \frac{1}{t^0} \log_e \left[\frac{V(t^0)}{C}\right] = \rho^0. \qquad \textbf{11}$$

The resulting t^0 will equal r^* only by coincidence. If t^0 and t^* are unique, we may assert that $t^0 = t^*$ if and only if $\rho^0 = r$.[7] This points to the source of the difference between the solutions of Boulding and, for example, Lutz and Lutz (see also Samuelson,

7. Proof of this proposition is supplied in the Appendix.

1937, pp. 469–96). Boulding apparently assumes that the periodic income given by expression **4** and occurring every t^* years may be spread over this interval in any desired pattern to suit the consumption of the investor by discounting at the internal rate of return ρ^0 rather than at the market rate r. If, as may be generally supposed, $\rho^0 \neq r$, then the period of investment Boulding derives will not be optimal.

Boulding's solution is still not optimal even in the situation in which a tree is planted at every instant, to be cut down after some interval of time t. If, however, the entire net proceeds from a mature tree can be reinvested in saplings (without any consumption), then after a period of time τ the total stock will be[8] $[V(t)/C]^{\tau/t}$. The logarithm of the rate of growth of this stock is

$$\frac{1}{\tau}\log\left[\frac{V(t)}{C}\right]^{\tau/t}, \qquad \textbf{12}$$

which reduces to the right side of **10**; hence the maximal rate of growth is given by maximization of ρ, i.e. by ρ^0.

In replacement problems, of course, it is a single chain which is of interest. Thus the opportunity cost of the firm's funds, usually called the cost of capital, is one of the necessary components of the analysis. The optimal life is obtained by minimizing the equivalent rate of expenditure, calculated by means of the cost of capital. Similarly, this rate of expenditure is also used to compare alternative replacements, possibly with different optimal lives.

Appendix

If $t^0 = t^*$, then $\rho^0 = r$, where t^0 corresponds to the maximum internal rate of return, ρ^0, and t^* is the optimum period resulting from maximization of the income stream $R(t)$ in expression **5**. From equation **11** we have

$$V'(t^*) = \rho^0 V(t^*) \qquad \textbf{13}$$

8. I.e., the first tree permits planting $V(t)/C$ new saplings which cycle is repeated τ/t times. This assumes, in addition, that the process can continue for a length of time without, for example, covering the globe (including oceans and deserts) with trees.

and from 3 $$V'(t^*) = \frac{r[V(t)^* - C]}{1 - e^{-rt^*}}. \qquad 14$$

Hence $$\rho^0 V(t^*) = \frac{r[V(t^*) - C]}{1 - e^{-rt^*}} \qquad 15$$

and $$C = \left[1 - \frac{\rho^0}{r}(1 - e^{-rt^*})\right] V(t^*). \qquad 16$$

By definition 9 $$C = V(t^*)e^{-\rho^0 t^*} \qquad 17$$

and so $$\frac{1}{\rho^0}(1 - e^{-\rho^0 t^*}) = \frac{1}{r}(1 - e^{-rt^*}) \qquad 18$$

or $$\frac{1}{Q(t^*, \rho^0)} = \frac{1}{Q(t^*, r)},$$

where $Q(t^*, r)$ is the 'annuity' (continuity) factor for interest rate r and an interval of length t^*. Therefore $\rho^0 = r$.

If $\rho^0 = r$ and t^0 and t^* are unique, then $t^0 = t^*$. We prove this by showing that $R(t^0)$ is a local maximum, hence satisfies the condition for t^*. Differentiating 5 we obtain at $t = t^0$,

$$R'(t^0) = \frac{r}{(e^{rt^0} - 1)^2}\{V'(t^0)(e^{rt^0} - 1) - re^{rt^0}[V(t^0) - C]\}. \qquad 19$$

Using 11 and the premise $\rho^0 = r$, we have

$$V'(t^0) = \rho^0 V(t^0) = rV(t^0), \qquad 20$$

and from definition 9 we obtain

$$C = V(t^0)e^{-rt^0}. \qquad 21$$

Substituting into 19 yields

$$R'(t^0) = \frac{r^2 V(t^0)}{(e^{rt^0} - 1)^2}\{(e^{rt^0} - 1) - (e^{rt^0} - 1)\} = 0.$$

References

BELLMAN, R. (1957), *Dynamic Programming*, Princeton University Press.
BOULDING, K. (1935), 'The theory of a single investment', *Q. J. Econ.*, vol. 49.
BOULDING, K. (1955), *Economic Analysis*, 3rd edn, Harper.
LUTZ, F., and LUTZ, V. (1951), *The Theory of Investment of the Firm*, Princeton University Press.
MORRIS, W. T. (1960), *Engineering Economy*, Irwin.
SAMUELSON, P. A. (1937), 'Some aspects of the pure theory of capital', *Q. J. Econ.*, vol. 51, no. 3.
TERBORGH, G. (1949), *Dynamic Equipment Policy*, McGraw-Hill.

8 T. L. Saaty

On Queueing History

Excerpt from T. L. Saaty, *Elements of Queueing Theory*, McGraw-Hill, 1961, p. 25.

We briefly mention the classification of queues due to D. G. Kendall. If M refers to the fact that the distribution of occurrences in a fixed time is Poisson (and hence that the distribution of the intervals is exponential), then $M/M/c$ refers to the c-channel case with Poisson input and exponential service time. The symbol D is used for the constant distribution, K_n for the chi square with even degrees of freedom, M_s for service in batches, E_n for the Erlangian with mean independent of n, GI for a general independent distribution (e.g. input intervals may be independently and identically distributed by a general distribution), and G refers to a general distribution without requiring independence. Thus $D/K_n/1$ refers to a constant-input, chi-square service-time distribution in the single-channel case, etc.

9 A. Lee

A Coupon Sorting Problem *and* Four Selected Case Histories

Excerpt from A. Lee, *Applied Queueing Theory*, Macmillan, 1966, pp. 136–51.

A Coupon Sorting Problem

This example concerns a single-server queue. Figure 1 illustrates the layout of a battery of common check-in counters at an airport. Passengers, indicated by crosses, joined on to queues before a number of counters. On arrival at a counter, a passenger handed over his ticket, his baggage was weighed and the weights entered on to the ticket, and the coupon for his immediate flight was taken from the ticket. The point of interest was what happened to his used flight coupon.

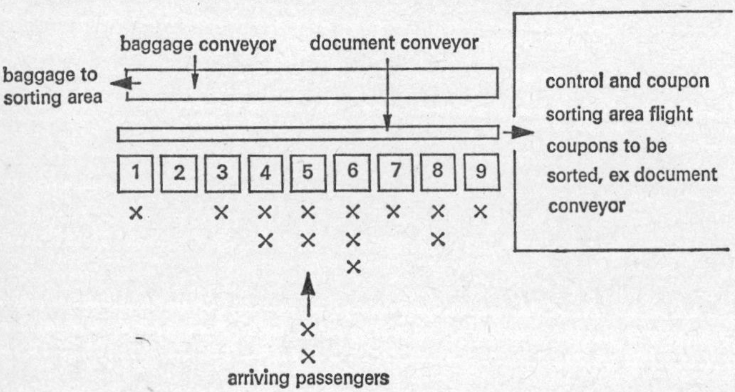

Figure 1 Sketch layout of coupon and baggage conveyors in relation to check-in counters

In fact the check-in agent put the passenger's used coupon into a document conveyor, in which it was carried to a sorting point. At this point, the coupon was released by the conveyor and fell freely into a bowl. A clerk took coupons from the bowl, one-by-one, inspected each and filed it in a pigeon-hole according to the flight number marked upon it. (The later history of the sorted coupons need not concern us.) Delays to the sorting of coupons might in turn produce delays to the departure of flights. It was, then, quite important that sorting delays should be short, but it was difficult to achieve this merely by doubling-up on sorters. Two sorting-clerks would tend to get in each other's way and this interference would impair their capability.

For this reason the question of the maximum sorting capacity of a single clerk was of interest. Indeed it might limit the number of counters which could feed into a single sorting point, with consequences for the design of passenger terminal check-in areas that could not be ignored. Ergonomic studies had already resulted in a highly efficient design for the sorting location, and there was no prospect of further reduction in the time taken to sort each coupon. The problem was one of calculation.

What was the capacity of a single sorting-clerk? Or more generally, what was the queueing-time distribution at different traffic intensities, of the single-server queueing-process in which: '"customers" were represented by coupons; "an arrival instant" was when a coupon fell into the bowl; the server was a sorting-clerk.' We must face up to the fact that to the airport manager this seemed a simple question to ask an operational research analyst. One can imagine him saying, 'If you can't tell me the answer to this little question, what use are you?' And of course in saying so he would be quite right. But when one comes to look at the queueing-process more carefully one finds oneself in strange country.

First of all, what about the input? Well, the number of coupons arriving at the sorting-point per hour (say) was equal to the number of passengers joining their flights during the same period (if a short lag is ignored). It is clear that the number of passengers checking in per hour was in some way related to the number of flight departures during some subsequent period. And it was well known that the flight departure schedule varied over the day. Consequently the average rate of coupon arrivals was not constant.

The reader, familiar with the ideas of busy periods, will say, and quite rightly, that this had nothing to do with the inputs involved in the maximum capacity problem. At maximum capacity the average input rate of coupons might be assumed steady over an extended period. That was so. There were however two secondary effects of varying input rates of some practical importance.

First; did the average sorting-time per coupon, taken by the clerk, vary with the rate of arrival of coupons? If not, perhaps all was fine and dandy. But suppose it did, what then? Were there signs of stabilization as the arrival rate increased? If so, perhaps the service-time corresponding to the maximum input rate could be computed. But again, suppose there were no signs of stabilization? Observation showed unequivocally that sorting-clerks worked faster when faced by a pile of unsorted coupons than when coupon arrivals were few and far between. It was happily, equally true that as the rate of arrivals increased, they tended to reach a minimum average sorting-time and not better it. (It is highly doubtful whether the sorting-rate represented by the lowest average could be sustained in real-life conditions. Each observation-period only lasted about fifteen minutes.) In the event another factor reared itself up during the tests that imposed a more stringent limitation on the average service-time: the proportion of coupons wrongly sorted. A wrong sort represents, say, an Amsterdam passenger credited to a Moscow flight and a great deal of subsequent confusion.

One of the supervisors at one airport where these problems were being studied suggested that the error-rate – the proportion of mis-sorts – increased during busy periods. This was rather difficult to establish in the real-life situation, and the O.R. team staged and carried out some special tests with the assistance of four sorting-clerks to check the truth of this. A definite threshold appeared above which errors became frequent.

As a result of these experiments and others, the team established that the lowest sorting-time per coupon that could be used in calculation was 2·1 seconds, with a coefficient of variation of $<0{\cdot}1$. In effect, therefore, the service-time distribution was constant.

Now let us return to the inputs. A casual analysis of arrival instants during short, busy periods, showed that the inter-arrival distribution was negative-exponential. We should however look at this result with some care: coupon arrivals at the sorting-point were generated by the departures of customers from a finite, and in point of fact quite small number of check-in counters: not exceeding eleven in the busiest periods.

During a busy period there was always someone waiting or being served at each check-in counter. The departures from each counter took place at instants separated by service-times. The inter-departure interval at each counter therefore had the identical distribution to the service-time distribution. This was of the Eulerian type. If we ignored the (relatively short) different lags between customer departures and coupon arrivals occasioned by the variable-length journey down the conveyor then the inter-arrival times at the sorting-point would have the same distribution as the pooled outputs of several renewal-processes with identical renewal-time distributions. It has been shown by Cox and Smith that under such circumstances the inter-event distribution of the pooled outputs is approximately negative-exponential. Thus some sort of rationale for the empirical result had been established.

This argument was not at all valid during slack periods. We were not, however, concerned with anything but peak periods. So far then we had (approximately) Poissonian arrivals, and a constant service-time. The model in Kendall's (short) code was $M/D/1$. 'Nothing to that' I almost hear the reader say. However Kendall's code does not include the queue-discipline. Each coupon fluttering from the conveyor fell into the bowl *on top* of the earlier arrivals, or sometimes due to some effect of the aerodynamics of falling pieces of paper, one would slide into the side of the untidy pile in the bowl. If the girls who took turns at this sorting job had merely taken coupons in their order of proximity to the top of the pile, the order of service would have been fairly near to a LIFO rule. But they were not so easily taken in, as previous experiences had shown how very long coupons at the bottom of the bowl might have to queue to be sorted during busy periods. So, they attempted to pull coupons from the bottom of the bowl first to maintain FIFO service. Unfortunately the bowl was rather slippery, as well as large in relation to the dimensions of the coupons, and it had steep sides so that occasionally coupons rested on the sides rather than the bottom. These factors conspired to make the actual queue discipline resemble SIRO rather than anything else. A few tests carried out by sending serially numbered coupons into the bowl from two of the conveyor input positions tended to confirm this. A SIRO rule was not absolutely correct as a description of the process, but it was considerably more realistic than a FIFO rule or a LIFO rule.

The queueing-process to be studied therefore could, it seemed, be represented by the model $M/D/1:(\infty/\text{SIRO})$. And the real point at issue was to be able to predict, for various ranges of input, the probability that a coupon might be delayed in the bowl for more than some critical length of time. The tolerable delay was, for reasons which will be evident to the reader of the previous chapter [not included here], set fairly low: not more than 10 per cent of coupons must be delayed for more ten seconds.

In the year that this problem was studied, Burke's charts for the model $M/D/1:(\infty/\text{SIRO})$ were not available. If they had been there would have been no difficulty. All that the team had available were charts for the following:

$$M/M/1:(\infty/\text{SIRO})$$
$$M/M/1:(\infty/\text{FIFO})$$
$$M/D/1:(\infty/\text{FIFO}).$$

And not too much time. (The problem did not merit a great expenditure of time and money. Its prompt solution was more a question of honour than anything else; there were much more important problems in process of being studied.)

If these three models are arranged in order with that which, at any traffic intensity, produces the fewest long queueing-times first, then we have

$$M/D/1:(\infty/\text{FIFO})$$
$$M/M/1:(\infty/\text{FIFO})$$
$$M/M/1:(\infty/\text{SIRO}).$$

It was certainly reasonable to assume that the model in which interest lay, $M/D/1:(\infty/\text{SIRO})$ would yield more long waiting-times than the first of these, but fewer than the last. For a single server, therefore, the maximum acceptable utilization to hold queueing-times within the prescribed limits must lie between:

80 per cent corresponding to the first model and 53 per cent corresponding to the third.

(Both results were read from charts and might have been subject to some reading error.)

The maximum input rate of coupons must therefore lie in the range of 910 (approx.) to 1370 (approx.) per hour. The rate of input of coupons to the sorting position was not likely to cross over into this range from below for a few years, and would certainly not, if the company's forecasts were even reasonably good, exceed the top of the range for five years. Thus the operational problem was resolved. The team had obtained an operational solution without solving – in the mathematical sense – the formal problem at all. In fact, this operational solution, in view of the questionable character of the queue-discipline was perhaps better than a mathematical solution of the apparently most appropriate model.

Subsequently, however, Burke's paper (1959) became available in the Bell System Technical Journal and was drawn to the attention of the O.R. group by the Communications Department of the company. A quick reference to Burke's charts showed that the maximum utilization given by the model $M/D/1:(\infty/\text{SIRO})$ was about 79 per cent – nearly at the top end of the range. In other words there was barely any difference between the two models $M/D/1:(\infty/\text{FIFO})$ and $M/D/1:(\infty/\text{SIRO})$ at utilizations near 80 per cent. Was this true? A further reference to Burke's paper, and to his Figure 4 (Burke, 1959), showed that this was indeed true.

It was comforting to know that the upper maximum input rate was the more nearly true, and that the sorting process could be operated by one person for up to (by then) three years ahead.

Four Selected Case Histories

Airport Taxiway Design

The first case concerns an investigation carried out at the end of the 1940s in England by mathematicians of the National

Physical Laboratory in conjunction with the Ministry of Civil Aviation. The problem was the design of the runway and taxiway systems for the new London Airport which was then in course of erection on part of what used to be known as Hounslow Heath. I cannot do better than quote verbatim an outline of this work provided by one of its leading participants, K. D. Tocher, who has written as follows:

The flow of material through an industrial plant is limited by the capacity of the process, handling and storing facilities available, and bottlenecks appear which slow the whole process down to the level of the slowest or smallest unit. [An] Operational Research Department is frequently asked to comment on this [type of] situation, and produce evidence of the effect of changes in plant capacity or changed methods of working.

The general results of Queue Theory establish that a mere evaluation of average throughput rates will be misleadingly optimistic. The question arises whether it is possible to design a control system which will reduce the loss of throughput caused by stochastic variation.

Queue Theory will rarely solve these problems completely, partly because of the intractable models necessary to be realistic, but more important, because it is a descriptive theory, predicting the effect of variability in the system. Once the queue discipline or the start of processing is dependent on the state of the plant (which is the effect of a control system), the mathematical models developed by Queue theorists are no longer useful. A theory of stochastic servo mechanisms is then required. No serious attempt seems to have been made in this field.

However, Queue Theory can often give useful indications about an industrial congestion problem.

[An example] is a very old application (some ten years old) concerning the design of an airport. The design of the runway and taxiway system at a major airport was under discussion, and the problem of congestion on the taxiways needed study. It was proposed that aircraft should taxi freely between the runways, the loading (and unloading) bays and the hangars, only subject to the rule that all aircraft making a given journey should follow a fixed route along the taxiways. Several taxiway systems had been proposed, and for each it was required to choose a routing system. The system which would handle the greatest throughput of aircraft and some indication of the level of this throughput were required (Tocher, 1961).

Tocher and his colleagues were attempting to tackle two types of operational problems: to *compare* some different systems, and to *estimate* some measure of effectiveness, or parameter, of one of them. We would be prepared to believe that the first problem could be quite satisfactorily resolved by the use of relatively crude models. The second we would expect to be more difficult. Tocher continues as follows:

Each taxiway system involved intersections at which aircraft could be held up. This was one source of congestion. Each bay had a limited capacity, and was subjected to the restraint that only one aircraft could move on the apron in front of the bays.

Thus the system could be modelled as a network of queues being fed from a variety of taxiways, processed at each intersection or bay, and whose outputs were routed along a choice of leaving taxiways. With general service and taxiway traverse times, this network gives rise to an insuperable mathematical problem, as the state of the queue at any one point depends upon the behaviour of queues at varying past times. However, if the system is assumed to be statistically stable, and only the equilibrium state be studied, some progress can be made.

The remaining assumption for this is that all process times (times to cross an intersection, taxiway traverse times, loading and unloading times) have negative exponential distributions. The whole problem then dissolves into a set of independent queues.

For each of these queues, the only information required is the total average rate of input and the mean holding time. Given the routing rules and the traffic densities on different routes, the input rates are formed as the solution of a set of linear equations. It is then possible to give the expected queue at each queueing point, and so present a bird's-eye view of the number of waiting aircraft at various points in the system at any random instant of time. A little elaboration enables the distribution of these quantities to be derived.

From the passenger's point of view, it is the total time and the number of delays encountered that are of interest. With the model, at the expense of some too heavy arithmetic, this could also be deduced.

The figures so obtained could not be accepted as absolute values, but they were adequate for comparative studies to eliminate certain of the proposed systems.

The determination of the maximum level of throughput was equally simple. Each individual queue must be stable, and so the problem reduced to studying the traffic densities which saturated the various queues. In this way the bottleneck intersections were revealed.

It was also possible to adopt the technique to design for a given taxiway system the best routing pattern for the various journeys. This established that one of the proposed routing systems was as good as could be devised (for the estimated relative traffic densities). The whole study revealed a disappointingly low throughput, and finally the uncontrolled use of the taxiways was abandoned and a traffic control system installed (Tocher, 1961).

The interesting feature of this problem is that as the best throughput that could be achieved with the best system was unacceptably small, one of the real problems that often occurs in queueing investigations was avoided. This is the problem of choice according to a criterion based upon queueing or waiting-times. If, say three systems had all turned out to permit an acceptable high throughput, then it would have been necessary to choose between them on the basis of which provided the lowest waiting-times, or even further which, whilst satisfying some waiting-time criterion, was cheapest. In other words, what Tocher calls the maximum level of throughput – the *saturation* throughput – would, it would seem reasonable to assume, be a great deal higher than the *critical* throughput which, in terms of waiting-times would be operationally acceptable and feasible.

The reader is asked to consider the possibility of the rankings of different queueing systems changing with changes in system parameters. This is a problem that did not occur in the study described by Tocher because of the disappointing best saturation throughput. This suggests that it may often be a good idea to begin an investigation of this type by computing saturation flows. By doing so a great deal of tedious and difficult work which would go for nothing may, from time to time, be avoided.

The London Coach Problem[1]

The second example has also to do with operations at London Airport, but in this instance the organization concerned was B.O.A.C. London Airport is not operated as most North American airports are, because it is not designed on the same *finger and gate* principle. In general aircraft which have just arrived, or which are about to depart, are located on 'stands' which are not always immediately adjacent to the terminal buildings. More often than not, therefore, coaches (buses) have

1. This selection is based upon the most interesting paper; see Friend (1958).

to be provided to transport passengers from the terminal to the aircraft or vice versa. About eight years ago B.O.A.C. undertook an investigation of the method of allocating coaches to this type of work, in so far as their own flights were concerned. The study was made by J. K. Friend, at that time of B.O.A.C.'s Operational Research Branch.

One of the methods of operation required the maintenance of a 'pool' of coaches on the apron (ramp, that is to say) to supply transport capacity at a moment's notice. As the arrival and departure times of flights are subject to some statistical variation around their scheduled times, for a great variety of reasons, the demand for coaches on the apron was a somewhat uncertain quantity, likely perhaps to exceed the supply from time to time. Friend's problem was therefore to compute the minimum number of coaches that should be kept in the pool at different times of day to ensure that on not more than a certain proportion of occasions would a demand occur at a time when all coaches were busy.

This might be thought of as a queueing problem. The coaches are service-channels, the arriving (or departing) flights are the customers and the service-times are those times for which coaches are busy with flights. They are *block-times*; that is, a service-time is the time for which the needs of one flight deny the use of a coach to another. The service-times were found to be approximately constant, with a mean of 7·5 minutes and a standard deviation of 1·7 minutes. A constant value was assumed to be valid.

The input to the queueing-process was, however, far from a Poisson process with constant mean. Analysis of historical data showed that the input during an hour period was very nearly Poisson with a mean value given by

$$m_r = \tfrac{1}{4}(s_{r-1} + 2s_r + s_{r+1})$$

where s_r is the scheduled movements during the rth hour, and a_r the actual movements.

During any hour, therefore, the process appeared to be such that the model $M/D/c$:(∞/FIFO) would be highly appropriate. Unfortunately the number of passengers to be transported per flight was a variable so that a proportion π_1 of flights required only one coach and a proportion π_2 required two ($\pi_1 + \pi_2 = 1$). The input consisted therefore of two separate Poisson inputs: one of single 'customers' at a rate $\pi_1 m_r$ and one of pairs of 'customers' at a rate $\pi_2 m_r$. The reader will quickly perceive that a model of the form $M(\pi_1 m, \pi_2 m)/D/c{:}(\infty/\mathrm{FIFO})$ is fairly discouraging, and will appreciate the difficulty of obtaining analytic solutions.

Friend devised an iterative computing procedure which he used to compute steady-state values of the cumulative probabilities of i or less customers being present (that is to say demand for coaches being satisfied or waiting to be met) simultaneously. These he used to construct a chart relating (a) the percentage of aircraft movements inadequately served (whether by lack of one each or two) to (b) the estimated number of aircraft movements per hour, for (c) various numbers of coaches from 2 to 10. The reader is referred to Friend's own account for details of the iterative computations.

The interest of this study as an illustrative example is the way Friend settled the question of predicting the rate of input by means of a weighted smoothing formula, and in doing so obtained an approximately Poisson steady input for periods of one hour.

Traffic Delays at Toll Booths

The most celebrated of all operational research studies of real problems involving queueing was carried out in the early 1950s by L. C. Edie of the Port of New York Authority, and later described in a paper[2] which won for him the first Lanchester Prize. Eleven years later the paper, and the study it describes remain classics with undiminished stature. This section is a very brief outline of Edie's work.

One of the key activities of the police personnel employed by

2. This is the most celebrated account of an operational research study of a non-telephone type queueing problem which has so far appeared. The reader's attention is drawn to the fact that in Edie's study, a relationship was established between mean service-time and queue-length (or mean arrival rate). Here then are three cases where one of the leading assumptions of classical queueing-theory (formulated under the predominant influence of telephone traffic engineering) namely independence of the service-time distribution – does not hold. It is a desirable experience to read Edie (1954).

the Port of New York Authority is the collection of tolls on bridges and in tunnels. In 1952 the payroll costs which this task generated were more than one million dollars per year. Preliminary analyses of toll booth operations revealed however that this expenditure was not being made to the best effect: the standard of service to patrons was very variable from one location to another and from one time of day to another, whilst the utilization of the traffic officers was also noticeably uneven.

The problem was to evaluate the standard of service given to customers at toll booths and find out how this varied with traffic volume in the toll lanes; to establish what standards of service should be offered; and finally to find some means of scheduling manpower most economically to provide these standards.

Several hundreds of man-hours of observation were expended to obtain empirical data which could be used to establish the most appropriate model to use in studying the process. After a further investment of several hundred man-hours in analysing these data, Edie and his associate managed to obtain a satisfactory model. The details of this phase of the project are to be found in Edie's paper: there is not space to reproduce them here.

The input process appeared to be Poissonian at traffic volumes around six hundred vehicles an hour, and Normal above that. The investigators concluded that this effect corresponded to the known degrees of closeness of approximation of the Poisson and normal distrubutions to the binomial distribution, and concluded further that the number of arrivals per unit time (30-second intervals) was distributed in the binomial form.

The service-time distribution was found to be constant. This result was established in two ways: first by taking measurements of some service-times; and second, by making comparisons between the actually observed curves relating queueing-times to traffic intensity for one booth with those given by the models $M/M/1:(\infty/\text{FIFO})$ and $M/D/1:(\infty/\text{FIFO})$. When the investigators came to look at multiple toll booth situations however, the observed delays were much greater than any given by the models $M/M/c:(\infty/\text{FIFO})$ and $M/D/c:(\infty/\text{FIFO})$. The reason was that in the multiple-channel case, the customers did not behave strictly according to the FIFO rule. Quite frequently a vehicle would drive up to join the queue at the end of one lane whilst another lane was empty. In other words this was not a *maître d'hôtel*, but a *ticket window without jockeying* system of queue formation. This would have been quite enough to explain the discrepancy between actuals and the theoretical results.

Edie also discovered that the service-time was related to the traffic volume. He then made use of the relationships which he had established between traffic intensity and delays on the one hand and service-times and traffic volume on the other, to develop curves relating average queueing-time per vehicle to traffic volume in vehicles per hour.

The next phase of the study was to determine the distribution of queue-lengths per toll lane. This was done by means which were generally similar to those employed in the previous phase. The importance of this characteristic of queueing-processes – which Edie calls *traffic backup* – in this particular situation was to identify how long delays to individual drivers might be, particularly bad perhaps in view of the lack of jockeying possibilities.

The next phase of the work, to which Edie devotes little attention in his paper unfortunately, was concerned with forecasting the hourly traffic volumes by day of the week. The traffic patterns for Tuesdays, Wednesdays, and Thursdays were found to be almost identical and could be combined. Other days needed individual treatment. It was found that the spread between median and peak traffic figures was of from 10 to 60 per cent at a bridge and 10 to 30 per cent at a tunnel.

Under these circumstances it was not, and could not be, possible to forecast the traffic load upon any particular toll plaza during any given hour on any chosen day with very great accuracy. With all due respect, and without in any way wishing to detract from Edie's achievement, the reader may wonder to what extent the vast amount of work carried out to determine a precise model of the queueing-process was necessary. For if two possible models produce results that differ by 10 per cent, and it is impossible to forecast the mean input rate to within a range of less than 50 per cent, the choice of model is likely to be a matter of slight significance.

Finally, the number of toll booths needed throughout a day was determined on the basis of optimum mean queueing-time for the median forecast traffic volumes, and calculated for half-hour intervals. These were then adjusted to reduce the peak levels of traffic backup to an acceptable level. Thus a schedule of toll booth requirements was obtained and it only remained to schedule manpower in such a way as to man such numbers of booths whilst providing for personal and meal relief periods. (Not much detail of this stage of the work is given by Edie.)

Appointment Systems[3]

One type of queueing-process is that in which the arrival-times of customers are scheduled in advance. Two examples of such processes are the arrivals of flights at an airport and the arrival of ore-ships at the wharfs of an ironworks. In both instances actual arrival-times differ from those scheduled, and this frequently makes it possible to approximate the real arrival pattern by a Poissonian one. But what if the customers do arrive at their scheduled times? This type of process has been studied in connexion with hospital out-patient appointment systems and has therefore come to be referred to as an appointment system or *appointment process.*

The models principally studied theoretically assume that the inter-arrival intervals, planned and actual, are of constant duration, so that the input process is regular. The analytical approach to such processes becomes quite difficult as soon as one moves away from the simpler assumptions concerning service-time distributions. Although the models such as $D/M/1:(\infty/\text{FIFO})$ are interesting, they are not very useful. Recently three case-histories have been published which describe how appointment systems have been studied. One of these, by Fry, is of unusual interest as it actually describes some empirical work – experimentation with a queueing-process rather than manipulation of a model. Another, by Welch, is concerned to some extent with the irresponsible behavior of the servers in such a system (in this case, doctors) who fail to commence work on schedule, and the effects of this behavior on the queueing-process. It too is extremely interesting. There are analogous situations in industry which do not appear to have been discussed at any length. For

3. For a fuller review, see Jackson, Welch and Fry (1964).

example, scheduled flights arrive at an airport to find that the landing runway is not available as it should be. It is the first article of the three which we shall look at in more detail, because this is a perfect little case-study of an operational research project.

The author of the article, R. R. P. Jackson, is a prominent queueing theorist. The problem which he discusses is to devise an efficient appointments system for a general practitioner. (It should be explained to non-British readers that the majority of general medical practitioners in Britain do not see their patients in surgery by appointment, and there is a more fundamental problem than the calculation of the optimum appointment system – namely to convince the doctors that they should have any sort of appointment system. For the benefit of British readers it should be explained that in Canada, and I believe in the U.S., it is usual to make appointments with general practitioners as well as with consultants.)

Jackson discusses a number of practical problems arising in the operation of an appointment system by a G.P.; for example the manning of the telephone in the doctor's office (again this is a peculiarly British problem, it appears) and the incidence of emergencies. He then explains that the queueing model appropriate to the situation is $D/E_k/1{:}(\infty/\mathrm{FIFO})$. In point of fact the assumption of infinite system capacity is not specifically stated, and there may have been a waiting-room assumption in his work. In this case the model would be $D/E_k/1{:}(\mathrm{L}/\mathrm{FIFO})$. In either case, existing analytical methods were inadequate for his purpose, and Jackson decided to simulate the model making use of a digital computer (and utilizing yet another simulation language, *Montecode*). This decision by so experienced an investigator is illustrative of the limitations of the theoretical, or analytical, approach to queueing problems.

As each surgery was limited to two hours duration (but one imagines the doctor would clear all patients present in his waiting-room at the close of the period) Jackson chose as the measure of efficiency for the doctor, or server, the total time spent waiting for patients. This would arise because occasionally the doctor would attend to a patient within the allotted inter-arrival interval whilst no patients with later appointments had yet arrived. By means of simulation runs he obtained estimates of the average patient waiting-time and the doctor's total waiting-time per surgery for various combinations of mean consulting-time (service-time) and inter-arrival time. These were charted and an optimum arrival interval obtained.

Jackson also investigated an approximation by comparing the simulated results for $D/E_k/1:(\infty/\text{FIFO})$ with those for $\text{D}/M/1:(\infty/\text{FIFO})$ obtained by letting $k \rightarrow \infty$ in the formulas for the model $E_k/M/1:(\infty/\text{FIFO})$ which he and Nickols had derived some time previously. He found that a traffic intensity of 0·91 in the case of $D/M/1:(\infty/\text{FIFO})$ was equivalent to one of about 0·81 in the case of negative exponential service-time and inter-arrival time distribution.

As a result of these studies, an optimum appointment system was devised and implemented by Jackson's own doctor, with advantage to everyone.

References

BURKE, P. J. (1959), 'Equilibrium delay distribution for one channel with constant holding time, Poisson input and random service', *Bell System Technical Journal*, vol. 38, no. 4.

EDIE, L. C. (1954), 'Traffic delays at toll booths', *Operations Research*, vol. 2, no. 2.

FRIEND, J. K. (1958), 'Two studies in airport congestion', *Operational Research Quarterly*, vol. 9, no. 3.

JACKSON, R. R. P., WELCH, J. D., and FRY, J. (1964), 'Appointments systems in hospitals and general practice', *Operational Research Quarterly*, vol. 15, no. 3.

TOCHER, K. D. (1961), 'The role of models in operational research', *J. Roy. Statist. Soc.*, part 2.

10 U. N. Bhat

Sixty Years of Queueing Theory

U. N. Bhat, 'Sixty years of queueing theory', *Management Science*, vol. 15 (1969), pp. B280–94.

In 'Seven more years of queues,' Saaty (1966) laments, 'in the past seven years the literature on Queueing Theory has increased by half of its amount for the previous fifty years. Improvements do not match the increase in theoretical developments. Rarely has so much ingenuity been shown in tackling a variety of technical problems on paper by some of the ablest people in the world. But real life queues are still primitive.' Prompted by these observations and the cynical remarks made by some of the practitioners in the field (see Lee, 1966, pp. 7, 27, 217), we would like to take stock of the present situation and try to discuss some of the reasons for the existence of such a state of affairs. We hope that a discussion of the problems, achievements, and the shortcomings of queueing theory will be helpful for both theorists and practitioners by putting things in proper perspective.

Looking at the literature on queueing theory in general we tend to agree with Dr Saaty's lament. The queueing concepts applied in practice do not show much indication that the applied scientists have gone much beyond the stage set in Morse's *Queues, Inventories and Maintenance* (1958). For instance, see a paper by Bhatia and Garg (1963) in which a mere description of and the problems connected with queueing systems is given. This is due, it seems to us, to the lack of communication between the so-called theoretical researcher and the applied scientist. (See Howard (1968) for a discussion of this problem in general.) Exceptions to these statements are some of the recent work done on optimization and simulation problems.

The next five sections of this discussion will be organized as follows. Section 1 gives the description of a queueing system as related to some practical situations and identifies the three problem areas as behavioral, statistical, and operational. In sections 2, 3, and 4, the literature on queueing theory is reviewed in the three specific problem areas mentioned above. Historical perspective is maintained throughout and comments are made on the directions taken by the subject and its consequences. Finally, we conclude by making some suggestions as to the desirable features in queueing theory research.

It should be emphasized that we envisage this study as purely descriptive, pointing out the type of work done in the development of the theory rather than saying what exactly has been done, which could be easily gathered by an interested researcher from the references cited in the discussion. It should also be noted that the references cited are directive rather than exhaustive.

Description and Problem Areas of a Queueing System

A 'queue' is a waiting line of units demanding service at a service facility (counter); the unit demanding service is called the 'customer' and the device at which or the person by whom it gets served is known as the 'server'. This terminology is used in a wide context. Here are a few realistic examples of this customer–server mechanism:

1. Vehicles demanding service arrive in a garage and depending on the number of employees, one or more vehicles may be repaired at a time.
2. Patients arrive at a doctor's clinic for treatment. Even if some appointment system exists, due to the emergency service rendered, there is a random element present in the arrival scheme and there is a possibility of a waiting line building up.
3. In a telephone exchange, the incoming calls are the customers who demand service in the form of telephone conversations.
4. Passengers demanding tickets queue up in front of a ticket counter.

It is possible to give numerous examples of this type, where a queue situation exists in one form or the other. As suggested by the above problems, some of these situations differ from each other in several details. However, it is not difficult to see that all these situations have certain common basic characteristics.

Input process. If the arrivals and service rates are strictly according to schedule, a queue can be avoided. But in practice this is not so and in most situations arrivals are controlled by external factors. Therefore, the best that can be done is to represent the input process in terms of random variables. Some of the factors needed for the complete specification of an input process are: the source of arrivals, the type of arrivals and the inter-arrival times.

Service mechanism. The uncertainties involved in the service mechanism are the number of servers, the number of customers getting served at any time and the duration of service. Again, random variable representations for these characteristics seem to be essential.

Queue discipline. All other factors regarding the rules of conduct of the queue can be pooled under this heading. One of these is the rule followed by the server in taking the customers in service. In this context, the rules such as 'first-come, first-served', 'last-come, first-served', and 'random selection for service' are self-explanatory. In many situations 'priority' disciplines need to be introduced, so as to make the system more realistic. Some of the other factors that may be included here are based on the customer behavior, such as balking, reneging, and jockeying.

Number of queues. When there is only one server in the system, specification of the above three factors gives a complete description. However, in problems of network flows and job-shop scheduling, one has to deal with more than one queue (server) in series and/or in parallel.

Based on these descriptions the problems arising in queueing theory can be classified into three.

Behavioral problems of the system

The study of behavioral problems aims at understanding a particular situation as thoroughly as possible. This is done by using mathematical models. Naturally, these are idealized models to varied degrees of realism. As done in many other branches of science, these models are studied analytically in isolation, with the hope that the information obtained from such a study will be useful in the decision-making process regarding such situations.

Major characteristics that need to be studied to understand the behavior of a queueing system are the queue length (number of customers waiting at time t), the waiting time (the time a new arrival will have to wait till his service commences) and the length of the busy period (the length of time when the server will be continuously busy). These factors are dependent on the input process, service mechanism, and the queue discipline, which are subject to uncertainties and hence, are better described as random variables. The queue length and the waiting time are stochastic processes (family of random variables indexed by a time parameter) whose behavior is given by transition distributions, whereas the busy period is a random variable whose distribution is of particular interest. Expected values and other moments of these distributions need to be obtained for a greater understanding of the processes involved.

Another behavioral distinction that can be made in these studies is between time-dependent (transient) and steady state behavior. The behavior of a stochastic process indexed by a time parameter generally depends on time. However, under certain conditions, after the system has been in operation for a sufficiently long time, it settles down to a behavior, which is independent of time. This is known as the steady state (equilibrium) behavior characterized by steady state distributions obtained from the time-dependent distributions by making time tend to infinity.

Statistical problems of the system

By statistical problems we mean the problems of the study of empirical data, estimation and tests of hypotheses regarding queue situations. Chronologically, the statistical study precedes the behavioral study as could be seen from some of the papers by Erlang (as reported in Brockmeyer *et al.* (1948), Molina (1927), and others). For an insight into the correct mathematical model, which could be studied analytically to derive its properties, a statistical study is very fundamental.

Behavioral results make several assumptions regarding the basic idealized model. Naturally, there should be a mechanism by which these assumptions could be verified. Starting with testing the goodness of fit for the arrival and servicing distributions, one would need to estimate the parameters involved or test hypotheses concerning the behavior of these systems. The important question as to whether a certain system is really in equilibrium at a certain point can also be answered by using suitable test procedures.

Operational problems of the system

Under this heading we include all problems that are inherent in the operation of a real life queueing system. Naturally some of these problems are statistical in nature. Others are mainly related to the design, control and the measurement of effectiveness of such systems.

Looking at the variety of queueing systems possible, it is only natural to assume that some models are more appropriate in certain contexts than the rest. Picking the right system for the right situation can be done either by designing a system according to the specifications or by choosing and controlling the parameters according to the needs.

From the point of view of an applied operations researcher, who is interested in solving specific real life problems, this aspect is more exciting than the other two, though he may see the need for the statistical study. He may even think that the behavioral study is purely academic. This point, though, is highly debatable. However, looking at the literature on queueing theory, it is clear that the major advances made are mostly on the behavioral aspect. There have been some statistical studies, but they are few. Operational studies are still fewer. With the advancement of knowledge one would expect that latest results would be used in studies dealing with decision problems. Unfortunately to a considerable extent this has not happened in queueing theory. Most of the results used have been relatively simple and it is quite clear that the applied researcher has not benefited from the advancement of knowledge on the basic model. There seems to exist a wide gulf between the so-called theoretician, who is very likely to be a mathematician, and the applied researcher. In the next three sections we shall review some of the work done in these aspects and then try to suggest some ways of bridging the gulf between the two types of researchers.

Behavioral Problems

Historically, Johannsen's article published in 1907 and reprinted (1910) seems to be the first paper on the subject. But the method used in this paper is not mathematically exact and therefore, from the point of view of exact treatment, the paper that has historic importance is A. K. Erlang's (1909). In this paper he lays the foundation for the place of Poisson (and hence, exponential) distribution in congestion theory. His papers written in the next twenty years contain some of the most important concepts and techniques; the notion of statistical equilibrium and the method of writing down balance of state equations (later called Chapman–Kolmogorov equations) are two such examples. Special mention should be made of his paper 'On the rational determination of the number of circuits' (see Brockmeyer *et al.*, 1948, p. 216), in which an optimization problem in congestion theory was tackled for the first time.

It should be noted that in Erlang's work, as well as the work done by others in the early thirties, the motivation has been the practical problem of congestion. See for instance, Molina (1927) and Fry (1928). During the next two decades several theoreticians became interested in these problems and developed more general models which could be used in more complex situations. Some of the authors with important contributions are Pollaczek, Khintchine, Palm, Kolmogorov, Crommelin and Volberg. A detailed account of the investigations made by these authors may be found in books by Syski (1960) and Saaty (1961).

In particular, noting the inadequacy of the equilibrium theory in many queue situations, Pollaczek (1934) began investigations of the behavior of the system during a finite interval. Since then he has done considerable work in the analytical behavioral study of queueing systems; see Pollaczek (1964). The trend towards the analytical study of the basic stochastic processes of the system has continued, and queueing theory has proved to be a fertile field for researchers who want to do fundamental research on stochastic processes involving mathematical models.

The processes involved are not simple and for the time-dependent analysis more sophisticated mathematical procedures are necessary. For instance, for the queue with Poisson arrivals and exponential service times, under statistical equilibrium the balance of state equations are simple and the limiting distribution of the queue size is obtained by recursive arguments and induction. But for the time-dependent solution the use of transforms is necessary. The first solution of this time-dependent problem was given by Bailey (1956) and Lederman and Reuter (1956). While Bailey used the method of generating functions for the differential equations, Lederman and Reuter used spectral theory in their solution. Later Laplace transforms have been used for the same problem, and it has been realized that generating functions and/or Laplace transforms form a useful technique in the solution of such difference–differential equations. Several systems have been studied by this method and for a complete reference on other methods we refer to the bibliographies given in some of the books cited earlier.

Other methods which rely on the heavy use of transforms are Takács equation method (1955) (here balance of state equations are written for the waiting time process giving an intro-differential equation, to be solved by the use of transforms), the supplementary variable techniques of Cox (1955), Keilson, and Kooharian (1960) (here the non-Markovian processes are rendered Markovian by describing the process with sufficient number of supplementary variables and operating on them; in addition to this, Keilson, also in his later investigations, uses first passage problems connected with sections of the process studied in isolation) and the techniques of recurrence relations and renewal theory used in varied forms by Gaver (1959), Takács (1962), Bhat (1964a), (1968b), Neuts (1966) and Cinlar (1967a and b) (here, recurrence relations for transitions in between successive regeneration points of the process are constructed and the entire analysis is built on these relations, often using renewal theory arguments). Analytically these methods are powerful and can be used in very complex situations as demonstrated in the papers mentioned above as well as those referred to in bibliographies on the subject. Transforms are also useful sometimes in identifying the behavior of the process. However, there is a serious disadvantage in these methods, which an analytically oriented researcher tries to neglect. Plainly speaking, the results, given in terms of transforms, very often with more than one argument, fail to make sense to an applied researcher. Numerical inversion of transformations (see Bellman *et al.* (1966), Gaver (1960) is an answer to this problem. But at this stage, the inversion methods are either not sophisticated enough to handle the more complex situations or do not appeal to the applied researcher. In simpler situations, means and variances could be deduced without much difficulty from the transforms. But, as the transforms get complicated, even these operations require extra dexterity in mathematical manipulations. This, it seems to us, has been the main reason for the gulf that exists between the theoretician and the applied scientist, even though so much ingenuity has been shown in tackling a variety of technical problems on paper by some of the ablest people in the world. As Kendall (1964) has aptly put it, 'much of the detail of the queue-theoretic scene has been obscured by the Laplacian curtain'. As a result, a researcher with not much sophistication in mathematical techniques takes the easiest and simplest way out of his complex problems by assuming steady state from the start of the operation or saying that all arrival or service time distributions can be approximated by an exponential distribution without loss of significant information. As we shall point out below, a realization of the inadequacy of the situation is slowly dawning on at least some of the people who are working on these problems.

A significant contribution in this direction was made by Kendall (1951 and 1953) when he demonstrated the use of regeneration point technique (imbedded Markov chain technique) in the classification of states and in deriving the equilibrium distribution of the Markov chain imbedded in the queue length processes of the (i) Poisson arrivals, general service time and single server; and (ii) general arrival, exponential service and multi-server queueing systems. His method bared the basic nature of the stochastic processes underlying the system instead of clouding it under some equations which are not amenable for easy interpretation.

Another first in this type of work is that of Champernowne (1956) who considered the simple queue with Poisson arrivals and exponential service time as a random walk and derived the time-dependent behavior of the queue length process in explicit form without resorting to transforms. Later, using combinatorial arguments Prabhu and Bhat (1963a), (1963b), and Bhat (1967), (1968c) have given the time-dependent behavior of more general systems, ((i) Poisson arrivals and general service, and (ii) Recurrent arrivals and exponential service) in explicit forms, which can be easily interpreted as probabilities of certain events in the process. This seems to us an essential step in putting the more advanced work in the easy reach of an applied scientist. Some work has already been carried out in using the results of Prabhu and Bhat (1963a) for numerical purposes. In a recent investigation Bhat and Mann (1967) have shown that the explicit expression for transition probabilities can be programmed to give tables for the time-dependent behavioral aspects of queueing systems.

Some of the recent work of Takács as given in his recent book (1967) is in the same spirit as above, in not using transforms to derive the behavior of queueing systems. He uses some generalizations of the ballot theorem in the fluctuation theory of random variables.

For general queueing systems, where very few assumptions are made with regard to the distributions involved in the process, there does not seem to be a way out of using transforms. See, for instance, Spitzer (1956) who uses the distribution of the maximum of partial sums of random variables. Nevertheless, at least in a wider class of special cases, the results could be made easily accessible by those who use them in practical problems. For a classified bibliography for general queueing systems, see Bhat (1968c).

Another area of research which is basically behavioral in nature but significant from the point of view of the applied scientist is the field of approximation. Considerable work on the approximations to the behavioral results for the general queueing systems with very few restrictions and when the traffic intensity is close to unity have been done by Kingman and others (for an extensive bibliography see Kingman (1964)). For instance, Kingman shows that under some considerably weak conditions, as the traffic intensity tends to one, the waiting time has an exponential distribution. An analytical study of this nature helps the applied scientist in using some of the simpler approximate results with a certain degree of confidence. In the same spirit is the work of Marshall (1966) who gives inequalities for some queueing theory results and some of the recent investigations by Prabhu and others who study queueing systems which are not in steady state; see Prabhu (1967) which also contains other references. (One of the referees has also pointed out that an article by D. P. Gaver, Jr., and a three-part article by G. F. Newell are due to appear soon in the *Journal of Applied Probability* on non-steady state queues.) These areas of approximations and non-steady state queues are still to be explored further, and, in future years, we can expect more people to be working on problems of this type. Incidentally, we may point out that these are examples where practical applications of the theory is benefited by the more advanced study of a theoretician.

The main purpose of the study of queueing systems is to understand real-life queue situations as well as possible. To achieve this goal numerous models have been studied and tested; in a short discussion of this nature recounting all these attempts is impossible. Keeping in mind the queue introduced by Erlang with Poisson arrivals, negative exponential service times and 'first-come, first-served' queue discipline the later variants can be approached in three ways.

Even though the Poisson assumptions is adequate enough approximation for a good number of arrival schemes, consideration of other schemes have widened the applicability of results found in queueing theory. For example, the general independent input process includes all other arrival processes which are independent in nature. Several studies have come out for systems which include such an input process in combination with exponential, Erlangian (gamma type) or Weighted–Sum–Erlangian service time distributions (see bibliography in Bhat (1968c); for the last service time distribution which is a variant of a bulk system (see Miller (1959) and Bhat (1964b)). Of these, the Weighted–Sum–Erlangian distribution has been shown to be general enough to be applicable in a large number of practical situations; see Luchak (1956). C. B. Winsten (1953) and A. Mercer (1960) have considered the case of scheduled but inexact arrivals.

Changing the nature of service is another way of generalizing the basic system. Service times with arbitrary distributions have been used in conjunction with exponential, Erlangian and Weighted–Sum–Erlangian inter-arrival time distributions. (For a bibliography see Bhat (1964b and 1968c) and Miller (1959).)

Relaxation of some of the assumptions of independence in the arrival and servicing processes brings the queueing system much closer to reality. A fair amount of work has been done in this direction, and systems with what may be called as 'first-order dependence' have been investigated. For the sake of convenience, such systems can be grouped under five headings as follows. Some of the major articles in each category are given in brackets:

(i) Systems with correlated arrivals (Chaudhry (1967), Runnenburg (1961), Pearce (1967)); (ii) Systems with dependent service times (Ali Khan and Gani (1968)); (iii) Systems with state-dependent arrivals (Haight (1957 and 1960), Ancker and Gafarian (1963a and b), Finch (1959)); (iv) Systems with state-dependent service (Conway and Maxwell (1962), Harris (1967)); (v) Systems with interdependent arrival and service processes (Neuts (1966) Cinlar (1967a and b), and Conolly (1968)).

For a discussion of these problems and other extensive bibliography reference can be made to Bhat (1968a). For a working bibliography on Markov Renewal Processes and Their Applications see Neuts (1968).

Ordered queues with the discipline 'first-come, first-served', though ideal in many situations, may not be realistic or practical. Some of the alternative disciplines are 'last-come, first-served', 'random selection for service' (see Saaty (1966) for a bibliography on these) or some sort of assignment of priorities for customers (for a bibliography see Jaiswal (1968)).

The study of series of queues forms the basis of network-flow theory and job-shop scheduling problems. We shall omit discussion of these areas as they themselves are broad enough to merit separate discussions. For bibliographies on these areas, see Beneš (1965) and Conway, Maxwell and Miller (1967).

Statistical Problems

The types of problems that arise in a statistical study are such as (a) identification of the appropriate model for a practical situation, (b) estimation of parameters from the empirical data, and (c) drawing inferences regarding future operations. For applying some of the behavioral results in real world situations, therefore, a statistical investigation is an essential feature. This has been demonstrated by the studies of Erlang (see Brockmeyer, *et al.* (1948), Molina (1927), Fry (1928), Bailey (1952 and 1955), Edie (1956b), to mention only a few). Before applying any of the results and drawing conclusions therefrom, some of the basic assumptions must be put to test. For example, in most of the studies inter-arrival times and service times are assumed to be independent random variables. This assumption needs to be verified before using such results. Similar is the problem of statistical equilibrium. In this context some of the general treatments of such topics given in books in statistical theory and in Cox and Lewis (1966) and Billingsley (1961) are useful. Cox and Lewis give methods of statistical analysis of series of events and demonstrate the use of these methods by numerical examples. Billingsley has developed the maximum likelihood theory for Markov processes which is an essential feature of some of the simpler queueing models. In addition to these general treatments, studies have also been made in specialized situations. N. T. J. Bailey (1952 and 1955) has carried out a 'quasi-empirical' investigation in which parameters are estimated from the data on hospital appointment systems, and new appointment schemes which incorporate better decision rules based on the estimated values are simulated. More theoretical problems of estimation have been tackled by Clarke (1957) who gives maximum likelihood estimates for the parameters of a system with Poisson arrivals, exponential service and a single server, Beneš (1957) who gives a sufficient set of statistics for a similar model with infinite number of servers, and Wolff (1965) who derives likelihood ratio tests and maximum likelihood estimators for simple queueing models. Another interesting study is that of Lilliefors (1966) who gives confidence intervals for the traffic intensity, expected queue length and other measures in systems with Erlangian arrival and service time distributions. Also see Cox (1964) for a comprehensive survey of some of these statistical problems.

The problem of identification of the appropriate model for a real life situation assumes the availability of such models. This assumption may not be true and hence the search for models goes on. For instance, for the inter-arrival or service times, negative-exponential distribution may not be the right one to be considered. Then one has to search for an appropriate distribution in the literature that would be a good fit. In the process, if one can also retain the simplicity of analysis, such a model would have wider use. From this point of view, distributions such as the hyper-exponential (see Morse (1958)) or the Erlangian have proved to be convenient. Luchak (1956) has investigated in detail the possibilities of characterizing the inter-arrival time or service time distribution by a weighted sum of Erlangian distributions whose means are multiples of the mean of a basic negative exponential distribution. He has also shown that a finite number of terms in the sum is a close enough approximation for a general situation and has demonstrated the use of the method of moments in the estimation of the weights and other parameters involved.

Operational Problems

The study of real systems is motivated by the objectives of improving its design, control and effectiveness. In the process of such a study, if the researcher finds the need for the behavioral study, he does so by empirical and analytical means. However, in queueing theory, behavioral study has not assumed this subsidiary role in its development. Invariably, the state of the art here is compared with the state of the art in inventory theory. In the latter class of problems, because of the non-stochastic nature of arrival of inventory which is a part of the decision rule, the basic structures of the underlying stochastic processes are relatively simple and a behavioral study does not become necessary. In queueing theory the underlying stochastic processes are much more complicated, especially under the several ramifications of the system. Therefore, even though the initial objectives were the same in both situations, describing the behavior of an existing system took precedence over the design and control of problems in the latter. The fact that more theoretically inclined mathematicians found interesting problems of stochastic processes in queueing theory only aggravated the situation. Complex processes, in general, do not yield simple results, and hence, the applied scientist who wishes to apply some of the behavioral studies has not been able to keep pace with the theoretical developments. This state of affairs has led to certain simplifying assumptions, such as the assumptions of random occurrence of events and statistical equilibrium, thus defeating the purpose of initial investigations.

The above discussion should not give the impression that the behavioral studies are unnecessary. Although more practical results are yet to be derived, these studies have contributed to a better understanding of the situations and have had direct influence on the solution of some of the design and control problems. In some situations they have given confidence in the use of simpler results in complex situations as good approximations (see Kingman (1964)).

Further, it should not be overlooked that there have been a number of studies aimed at improving the system. Greater momentum is being gained by such studies during recent years and it is our intention here to mention a few of these.

As mentioned earlier, Erlang's interest in the subject was in terms of building better telephone systems for the company he was working for. His paper 'On the rational determination of the number of circuits' (Erlang, 1924) deals with determining the optimum number of channels so as to reduce the probability of loss in the system.

Another study, which has also been mentioned earlier in connexion with the estimation problem, is Bailey's (1952) in which a simulation procedure is adopted to get better appointment systems in hospitals. A similar study is Edie's (1956a and 1956b), which looks into traffic delays at toll booths. Cox and Smith (1961) and Benson and Cox (1951, 1952) use the slightly different approach of giving some observational characteristics which indicate the improvement in the system. They consider the productivity of machines requiring attention and give measures for the operative efficiency (proportion of time the repairman is working) and machine availability (proportion of time the machine is running) and estimate the loss of production for the system. Tables giving such information can be easily used in improving the system design.

In solving many of the complex operational problems, simulation has been a useful tool. Basically, the function of simulation is diagnostic rather than remedial. As long as the practitioner understands this limitation, it can be effectively used in improving the system. For some insight into these techniques see Tocher (1963), Lee (1966), and Page (1965).

From the point of view of applications of queueing results to realistic problems Morse's book (1958) has been held in high regard. This is due to the fact that he presents some of the theoretical results in a manner appealing to the applied researcher. Also, he gives some procedures of improving the system design. In a problem of balancing mean wait and service cost in a realistic situation in which ships arrive at the harbor for unloading, the main objective is to decide upon the optimum traffic intensity so that the advantages of service are not lost in waiting. In another problem of optimizing the number of service channels, the intention is to maximize the net profit for given values of arrivals rate, service rate in each channel, average gross profit per customer served and cost of service per channel per unit time. Also, the chapters on inventory theory and maintenance problems in his book are significant from the point of view of a research worker on queues because of the wide applicability of his results. (Also see Riordan (1962).)

Hillier's paper (1963) on economic models for industrial waiting line problems introduces standard optimization techniques to queueing problems. Using steady state characteristics of the system and cost functions of the time waited and the number of servers, he studies the problem of determining the parameters (arrival rate, service rate and the number of customers) in the multi-server system with Poisson arrivals, and exponential service (also see references cited in this paper). Another study of an optimization problem through classical means is that of Heyman (1968) who derives an optimal policy for turning the server on and off in the queue with Poisson arrivals and general service times.

Inventory theory has developed some powerful methods for deriving optimal policies for its problems. Attempts are being made to extend some of these methods to queueing theory and allied fields. In this connexion two papers by Bather (1962 and

1963) are significant. In these he discusses optimal time-dependent and equilibrium policies for the release of water from finite dams. (Dam processes are similar to the waiting time processes of queueing systems.)

The basic characteristics of a queueing system includes, apart from the arrival and service time distributions and the number of servers, the significant factor of queue discipline. Improvements for the design can also be made by a proper selection of the queue discipline. Optimization studies regarding these problems are therefore very important and during the last few years some attempts have been made in this direction. For instance, Brosh and Naor (1963) consider attaching priorities for different arrival streams (assumed Poisson) at a counter when cost functions are attached to the expected queue sizes in these different streams, the total cost being linear. They give general criteria for optimal priority ordering which minimize the cost under study. A similar problem has been studied by Oliver and Pestalozzi (1965) using the theory of dynamic programming. The objective function in this study is taken to be the expected queueing time of all customers using the service system.

Another type of optimization problem in priority queues has been investigated by Etschmaier (1966) in a single server queueing system with two priority classes. These classes have pre-assigned priorities and the discipline is such that a low priority customer is pre-empted upon the arrival of a high priority customer only if the elapsed service time is less than a number z. Optimal values of z are obtained for linear cost functions. From these considerations he and Jaiswal (1968) are able to consider the different discretionary priority rules (such as 'pre-emptive', 'head-of-the-line', 'first-come, first-served', etc.) as belonging to a general class of problems which can be defined by the use of (z, d) where z is the optimum value for the elapsed service time, as defined in the problem above, and d is the ratio of a measure of a factor of waiting time and an urgency number for the class of the customer (the urgency number of the low priority customer being equal to zero). Then the design problem centers around the determination of these numbers.

When arrivals form a Poisson process and service time distribution is negative exponential, some of the design and control problems connected with such systems fall under the category of Markov decision processes. Lately these problems have attracted considerable attention and some of the articles we may make reference to are by Miller (1967), Shaler (1968), Kumin (1968), and Evans (1968a, 1968b, and 1968c). Of these, H. Kumin and R. V. Evans have directed their efforts to studying the problem of system design by deriving a matrix of transition rates that are optimal.

As mentioned earlier, we do not intend to discuss problems in the theory of network flows and the theory of scheduling in this paper. Considerable work on optimization and other design mprovement problems have been carried out in these areas and those interested are referred to the recent books by V. Beneš (1965) and Conway, Maxwell and Miller (1967). From the Queueing Theory point of view, other authors in this field are Evans (1964 and 1967), Jackson (1962), and Kleinrock (1965), to name only a few.

Conclusion

A review of all the important developments of queueing theory in these six decades of its existence is not possible in a short paper. We have attempted to point out some of the major problems and the directions the investigations of these problems have taken. Looking into the three problem areas in the subject and recounting the important trends in these areas, we are able to get a better picture of the whole. Though it is true that work of 'practical' interest has lagged behind that of 'theoretical' interest, this is only to be expected and not entirely unhealthy. As pointed out earlier, the basic structure of the underlying processes are complex by comparison and it requires extra sophistication on the part of the researcher. Therefore, it has taken the expertise of some of the 'ablest people in the world' to give satisfactory answers to these problems. And it is always the case that the more one is theoretically oriented, the less he has the patience for details. This has widened the gap between the two kinds of work. A question can now be asked: What can be done to remedy this situation?

First of all, it is our firm belief that the applied researcher should give up the naïve viewpoint that the steady state behavior of a Poisson arrival, exponential service queueing system is all that needs to be considered in real life situations. Indiscriminate use of expected values strikes at the root of stochastic assumptions. The Poisson assumption has as much truth in it as the usual 'normal' assumption for distributions in statistical theory. As regards the steady state behavior an example would clarify the situation. One of the characteristics that has been studied numerically by Bhat and Mann (1967) for the single server queue with Poisson arrivals and constant service times is what can be termed as the 'first entrance time for steady state' (effective steady state). For instance, with the traffic intensity 0·7 it is found that it takes about forty departures after the start of the system to attain effective steady state, the result being accurate up to four decimal places. In an eight-hour work day with a service time of six or seven minutes duration, it takes more than half the day to come to a state where equilibrium results would hold. Further, the probabilities of finding the system empty can be easily seen to be monotonically decreasing, and hence, if one uses the smaller probability in design problems of the system, it is not hard to realize that there is bound to be some wastage of resources in the entire design. It must be pointed out again in this respect that queueing problems are different from inventory problems. In the latter case stochastic nature persists only in the demand process. With a periodic review the inventory process starts with every period and only the total demand in a period needs to be known. Whereas, for a queueing system, in a single period of a corresponding inventory system, both the arrivals and the services are to be accounted for one by one all through the period. Further, a day starts with a new queue which takes a good part of the day to settle down for a stable behavior whereas the inventory is continued from period to period without effecting the nature of the process. All these arguments show that what we need in the solution of design problems in queueing systems are not simplifying assumptions that reduce the problems into familiar patterns. Instead, what we need are more sophisticated procedures which can take care of the complexities of the situation. Progress in this direction has been slow indeed. We earnestly hope that more and more people would come forward to accept this challenge.

There are some people who take refuge in the blanket statement that 'queueing theory is closed' – meaning that nothing more interesting can be done in the subject. In fact we feel that the opposite is true. From the design point of view it has just started. More developments are yet to come. With the ever increasing demands for solutions for the design problems in computers and networks, new methods will be developed, new horizons will be drawn, and new dimensions will be conquered. The basic requirements for these things to happen are the right attitude and the right environment. This hope is reinforced by some of the recent developments in the subject. As mentioned earlier, the design and control problems of queueing theory are looked upon as Markov decision problems and advances in this area are clearly taking place in several centers around the country, and more and more people are using non-linear and dynamic programming methods in queueing problems.

At the same time, the need for behavioral and statistical studies also exists. One of the aims of the study of systems is to understand real life situations. In this respect, the need for relaxing some of the restrictive assumptions regarding the input process, servicing process and queue disciplines still remains. Study of such systems would shed more light on their behavior. A trend in this direction is already seen in papers dealing with dependent processes. The same is true about statistical analysis. More powerful techniques need to be developed to match theoretical results with the empirical data. For instance, assumption of equilibrium behavior in place of transient behavior is quite common. In such cases an investigation into the error involved is quite proper and essential. Another area would be to investigate the usefulness of the latest methods in estimation and testing of hypothesis in queueing problems.

An area that should not be overlooked is that of making the existing results more meaningful for the practitioner. In this respect more work needs to be done in the problem of numerical inversion of transforms and in the simplification of some of the complex closed form expressions, either by approximations or by putting them in convenient forms for computer programming. In this regard, see Leese and Boyd (1966) and Bhat and Mann (1967).

With these and many more significant unanswered questions we cannot conclude that there is not much by way of research in queueing theory. Looking back at the developments of the theory, we do not feel that there is any need for lament either.

References

ANCKER, C. J., JR, and GAFARIAN, A. V. (1963a), 'Some queueing problems with balking and reneging, I', *Operations Research*, vol. 11, pp. 88–100.

ANCKER, C. J., JR, and GAFARIAN, A. V. (1963b), 'Some queueing problems with balking and reneging, II', *Operations Research*, vol. 11, pp. 928–37.

BAILEY, N. T. J. (1952), 'Study of queues and appointment systems in out-patient departments with special reference to waiting times', *J. Roy. Stat. Soc., B*, vol. 14, pp. 185–99.

BAILEY, N. T. J. (1955), 'A note on equalizing the mean waiting time of successive customers in a finite queue', *J. Roy. Stat. Soc., B*, vol. 17, pp. 262–3.

BAILEY, N. T. J. (1956), 'A continuous time treatment of a simple queue using generating functions', *J. Roy. Stat. Soc., B*, pp. 288–91.

BATHER, J. A. (1962), 'Optimal regulation policies for finite dam', *J. Soc. Ind. App. Math.*, vol. 10, pp. 395–423.

BATHER, J. A. (1963), 'The optimal regulation of dams in continuous time', *J. Soc. Ind. App. Math.*, vol. 11, pp. 33–63.

BELLMAN, R. E., KALABA, R. E., and LOCKETT, J. A. (1966), *Numerical inversion of Laplace transform: Application to Biology, Economics, Engineering and Physics*, American Elsevier Publishing Co.

BENEŚ, V. E. (1957), 'A sufficient set of statistics for a simple telephone exchange model', *Bell System Tech. J.*, vol. 36, pp. 939–64.

BENEŠ, V. E. (1965), *Mathematical theory of connecting networks and telephone traffic*, Academic Press.

BENSON, F., and COX, D. R. (1951), (1952), 'The productivity of machines requiring attention at random intervals', *J. Roy. Stat. Soc., B*, vol. 13, pp. 65–82; *B*, vol. 14, pp. 200–219.

BHAT, U. NARAYAN (1964a), 'On single-server bulk queueing processes with binomial input', *Operations Research*, vol. 12, pp. 527–33.

BHAT, U. NARAYAN (1964b), 'Imbedded Markov chain analysis of single-server bulk queues', *J. Aust. Math. Soc.*, vol. 4(2), pp. 244–63.

BHAT, U. NARAYAN (1967), 'Some explicit results for the queue GI/M/1 with Group Service', *Sankhya A*, vol. 29(2), pp. 199–206.

BHAT, U. NARAYAN (1968a), 'Queueing systems with first-order dependence', Department of Operations Research, Case Western Reserve University, Technical Memorandum no. 129.

BHAT, U. NARAYAN (1968b), 'Transient behavior of multi-server queues with recurrent input and exponential service times', *J. App. Prob.*, vol. 5(1), pp. 158–68.

BHAT, U. NARAYAN (1968c), 'A study of the queueing systems', *Lecture Notes in Operations Research and Mathematical Economics*, M/G1 and GI/M/1, no. 2, Springer-Verlag.

BHAT, U. NARAYAN, and MANN, S. H. (1967), 'A numerical method for constructing tables for the behavior of some infinite queueing systems', Technical Memorandum no. 82, Operations Research Group, Case Institute of Technology.

BHATIA, A., and GARG, A. (1963), 'Basic structure of queueing problems', *J. Ind. Eng.*, vol. 14(1), pp. 13–17.

BILLINGSLEY, P. (1961), *Statistical Inference for Markov Processes*, University of Chicago Press.

BROCKMEYER, E., HALSTROM, H. L., and JENSEN, A. (1948), *The Life and Works of A. K. Erlang*, translation of the Danish Academy of Science, no. 2.

BROSH, I., and NAOR, P. (1963), 'On optimal disciplines in priority queueing', *Bull. Inst. Stat. Inst.*, vol. 40(2), pp. 593–607.

CHAMPERNOWNE, D. G. (1956), 'An elementary method of solution of the queueing problem with a single-server and a constant parameter', *J. Roy. Stat. Soc., B*, vol. 18, pp. 125–8.

CHAUDHRY, M. L. (1967), 'Queueing problems with correlated arrivals and service through parallel channels', *Canad. Oper. Res. Soc. J.*, vol. 5(1), pp. 35–46.

CINLAR, E. (1967a), 'Time dependence of queues with semi-Markovian services', *Journal of Applied Probability*, vol. 4, pp. 356–64.

CINLAR, E. (1967b), 'Queues with semi-Markovian arrivals', *Journal of Applied Probability*, vol. 4, pp. 365–79.

CLARKE, A. B. (1957), 'Maximum likelihood estimates in a simple queue', *Ann. Math. Stat.*, vol. 28, pp. 1036–40.

CONOLLY, B. W. (1968), 'The waiting time process for a certain correlated queue', *Operations Research*, vol. 16.

CONWAY, R. W., and MAXWELL, W. L. (1962), 'A queueing model with state dependent service rates', *J. Indust. Eng.*, vol. 12, pp. 132–6.

CONWAY, R. W., MAXWELL, W. L., and MILLER, L. W. (1967), *Theory of Scheduling*, Addison-Wesley.

COX, D. R. (1955), 'The analysis of non-Markovian stochastic processes by the inclusion of supplementary variables', *Proc. Comb. Phil. Soc.*, vol. 51, pp. 433–41.

COX, D. R. (1964), 'Some problems of statistical analysis connected with congestion', *Proc. of the Symposium on Congestion Theory*, University of North Carolina, Chapel Hill, pp. 289–316.

COX, D. R., and SMITH, W. L. (1961), *Queues*, Methuen Monographs on Applied Probability and Statistics, Wiley.

COX, D. R., and LEWIS, P. A. W. (1966), *The Statistical Analysis of Series of Events*, Methuen.

CRABILL, T. B. (1968), 'Optimal control of a queue with variable service rates', Ph. D. Dissertation, Cornell University.

EDIE, L. C. (1956a), 'Traffic delays at toll booths', *J. Operations Research*, vol. 4, pp. 107–38.

EDIE, L. C. (1956b), 'Optimization of traffic delays at toll booths', *Operations Research for Management*, vol. 2, ch. 3, Johns Hopkins Press.

ERLANG, A. K. (1909), 'The theory of probabilities and telephone conversations', *Nyt tiddskrift for matematik, B*, vol. 20, p. 33.

ERLANG, A. K. (1924), 'On the rational determination of the number of circuits', in E. Brockmeyer *et al.*, *The Life and Works of A. K. Erlang*, translation of ths Danish Academy of Science, no. 2, 1948, pp. 102, 216.

ETSCHMAIER, M. (1966), 'Discretionary priority processes', *Technical Memorandum*, no. 60, Operations Research Group, Case Institute of Technology.

EVANS, R. V. (1964), 'Queueing when jobs require several services which need not be sequenced', *Management Science*, vol. 10(2), pp. 298–315.

EVANS, R. V. (1967), 'Geometric distribution in some two-dimensional queueing systems', *Operations Research*, vol. 15(5), pp. 830–46.

EVANS, R. V. (1968a), 'Programming in Markov processes', *Technical Memorandum*, no. 123, Operations Research Department, Case Western Reserve University.

EVANS, R. V. (1968b), 'Programming problems and changes in the stable behavior of a class of Markov chains', *Technical Memorandum*, no. 126, Operations Research Department, Case Western Reserve University.

EVANS, R. V. (1968c), 'Design of discrete parameter Markov systems', *Technical Memorandum*, no. 127, Operations Research Department, Case Western Reserve University.

FINCH, P. D. (1959), 'The output process of the queueing system M/G/1', *J. Roy. Stat. Soc., B*, vol. 21, pp. 375–80.

FRY, T. C. (1928), *Probability and its Engineering Uses*, Van Nostrand.

GAVER, D. P., JR (1959), 'Imbedded Markov chain analysis of a waiting line process in continuous time', *Ann. Math. Stat.*, vol. 30, pp. 698–720.

GAVER, D. P., JR (1960), 'Observing stochastic processes and approximate transform inversion', *Operations Research*, vol. 14, pp. 444–59.

HAIGHT, F. A. (1957), 'Queueing with balking, I', *Biometrika*, vol. 46, pp. 360–69.

HAIGHT, F. A. (1960), 'Queueing with balking, II', *Biometrika*, vol. 47, pp. 285–96.

HARRIS, CARL M. (1967), 'Queues with state dependent stochastic service rates', *Operations Research*, vol. 15, pp. 117–30.

HEYMAN, DANIEL P. (1968), 'Optimal operating policies for M/G/1 queueing systems', *Operations Research*, vol. 16, pp. 362–82.

HILLIER, F. S. (1963), 'Economic models for industrial waiting line problems', *Management Science*, vol. 10, pp. 119–30.

HOWARD, R. A. (1968), 'The practicality gap', *Management Science*, vol. 14(7), pp. 503–7.

JACKSON, J. R. (1962), 'Networks of queues with controlled customer arrival rates', *Research Report*, no. 80, University of California, Los Angeles, California.

JOHANNSEN, Fr. (1910), 'Waiting times and numbers of calls', *Post Office Electrical Engineers Journal*, London, October and January.

JAISWAL, N. K. (1968), *Priority Queues*, Academic Press.

KEILSON, J., and KOOHARIAN, A. (1960), 'On time-dependent queueing processes', *Ann. Math. Stat.*, vol. 31, pp. 104–12.

KENDALL, D. G. (1951), 'Some problems in the theory of queues', *J. Roy. Stat. Soc., B*, vol. 13, pp. 151–85.

KENDALL, D. G. (1953), 'Stochastic processes occurring in the theory of queues and their analysis by the method of imbedded Markov chains', *Ann. Math. Stat.*, vol. 24, pp. 338–54.

KENDALL, D. G. (1964), 'Some recent work and further problems in the theory of queues', *Theory of Probability* (Russian), vol. 9, pp. 1–15.

ALI KHAN, M. S., and GANI, J. (1968), 'Infinite dams with inputs forming a Markov chain', *J. App. Prob.*, vol. 5(1), pp. 72–83.

KINGMAN, J. F. C. (1964), 'The heavy traffic approximation in the theory of queues', *Proceedings of the Symposium on Congestion Theory*, University of North Carolina, pp. 137–69.

KLEINROCK, L. (1965), 'A conservation law for a wide class of queueing disciplines', *Naval Res. Logistics Q.*, vol. 12(2), pp. 181–92.

KUMIN, HILLEL J. (1968), 'The design of Markovian Congestion Systems', *Technical Memorandum*, no. 115, Operations Research Department, Case Western Reserve University.

LEDERMAN, W., and REUTER, G. E. (1956), 'Spectral theory for the differential equations of simple birth and death process', *Phil. Trans. Roy. Soc.*, London, A, vol. 246, pp. 321–69.

LEE, A. M. (1966), *Applied Queueing Theory*, St Martin's Press.

LEESE, E. L., and BOYD, D. F. (1966), 'Numerical methods of determining the transient behavior of queues with variable arrival notes', *Jour. Canad. Oper. Res. Soc.*, vol. 4(1), pp. 1–13.

LILLIEFORS, H. W. (1966), 'Some confidence intervals for queues', *Operations Research*, vol. 14(4), pp. 723–7.

LUCHAK, G. (1956), 'The solution of the single channel queueing equation characterized by a time-dependent Poisson distributed arrival rate and a general class of holding times', *Operations Research*, vol. 4, pp. 711–32.

MARSHALL, K. T. (1966), 'Some inequalities for single-server queues', *Report*, no. 66–19, Operations Research Center, University of California, Berkeley.

MERCER, A. (1960), 'A queueing problem in which the arrival times of the customers are scheduled', *J. Roy. Stat. Soc., B*, vol. 22, pp. 108–13.

MILLER, R. G., JR (1959), 'A contribution to the theory of bulk queues', *J. Roy. Stat. Soc., B*, vol. 21, pp. 320–37.

MILLER, B. L. (1967), 'Finite state continuous time Markov decision processes with applications to a class of optimization problems in queueing theory', *Technical Report*, no. 15, Department of Operations Research, Stanford University.

MOLINA, E. C. (1927), 'Application of the theory of probability to telephone trunking problems', *Bell System Tech. J.*, vol. 6, pp. 461–94.

MORSE, P. M. (1958), *Queues, Inventories and Maintenance*, Wiley.

NEUTS, M. F. (1966), 'The single server queue with Poisson input and semi-Markov service times', *Journal of Applied Probability*, vol. 3, pp. 202–30.

NEUTS, M. F. (1968), 'A working bibliography on Markov renewal processes and their applications', Mimeo Ser., no. 140, Department of Statistics, Purdue University.

OLIVER, R. M., and PESTALOZZI, G. (1965), 'On a problem of optimum priority classification', *J. Soc. Indust. Appl. Math.*, vol. 13(3), pp. 580–91.

PAGE, E. S. (1965), 'On Monte Carlo methods in congestion problems', *Operations Research*, vol. 13, pp. 291–306.

PEARCE, C. (1967), 'An imbedded chain approach to a queue with moving average input', *Operations Research*, vol. 15, pp. 1117–30.

POLLACZEK, F. (1934), 'Über das warteproblem', *Math Z.*, vol. 38, pp. 492–537.

POLLACZEK, F. (1964), 'Concerning an analytic method for the treatment of queueing problems', *Proceedings of the Symposium on Congestion Theory*, University of North Carolina, pp. 1–42.

PRABHU, N. U., and BHAT, U. N. (1963a), 'Further results for the queue with Poisson arrivals', *Operations Research*, vol. 11, pp. 380–6.

PRABHU, N. U., and BHAT, U. N. (1963b), 'Some first passage problems and their application to queues', *Sankhya A*, vol. 25, pp. 281–92.

PRABHU, N. U. (1967), 'Queueing systems not in steady state', Paper presented in the Department of Industrial and Systems Engineering, University of Florida (private communication).

RIORDAN, J. (1962), *Stochastic Service Systems*, Wiley.

RUNNENBURG, J. TH. (1961), 'An example illustrating the possibilities of renewal theory and waiting time theory for Markov dependent arrival intervals', Proceedings Series A (Math. Sc.), *Koninklijke Nederlandse Akademia Van Wetenschappen*, vol. 64, pp. 560–76.

SAATY, T. L. (1961), *Elements of Queueing Theory with Applications*, McGraw-Hill.

SAATY, T. L. (1966), 'Seven more years of queues: a lament and a bibliography', *Naval Res. Logistics Q.*, vol. 13(4), 447–76.

SHALER, S., JR (1968), 'Static decision models for queueing systems with non-linear waiting costs', *Technical Report*, no. 9, Department of Operations Research, Stanford University.

SPITZER, F. (1956), 'A combinatorial lemma and its applications to probability theory', *Trans. Am. Math. Soc.*, vol. 82, pp. 323–39.

SYSKI, R. (1960), *Introduction to Congestion Theory in Telephone Systems*, Oliver and Boyd.

TAKÁCS, L. (1955), 'Investigations of waiting time problems by reduction to Markov processes', *Acta Math. Acad. Sci. Hung.*, vol. 6, pp. 101–29.

TAKÁCS, L. (1962), *Introduction to the Theory of Queues*, Oxford University Press.

TAKÁCS, L. (1967), *Combinatorial Methods in the Theory of Stochastic Processes*, Wiley.

TOCHER, K. D. (1963), *The Art of Simulation*, English University Press.

WINSTEN, C. B. (1953), 'Geometric distributions in the theory of queues', *J. Roy. Stat. Soc., B*, vol. 21(1).

WOLFF, R. W. (1965), 'Problems of statistical inference for birth and death queueing models', *Operations Research*, vol. 13, pp. 343–57.

11 C. C. Pegels

On Startup or Learning Curves: An Expanded View

C. C. Pegels, 'On startup or learning curves: an expanded view', *A.I.I.E. Transactions*, vol. 1 (1969), no. 2, pp. 216–22.

Introduction

'Learning curves', 'startup curves', 'cost reduction curves', or by whatever other name they are known, have many potential applications in the areas of production planning, budgeting, purchasing, and contract bid preparation. A review of 'learning curves' applications by Baloff (1967) indicates that the power function formulation can be used to describe the productivity increases that accompany the introductions of new products in a variety of labor-intensive assembly situations, including the manufacture of airframes, electronic and electro-mechanical components, and machine tools. In addition, the model has also found applicability to instances of both product and process startup in the machine-intensive manufacture of steel, glass, paper, and electrical products.

The literature so far has concentrated mainly on a somewhat standardized power function formulation. This formulation was first introduced by Wright (1936). Since then this power function formulation has managed to survive in essentially its original form, even though it has many basic weaknesses.

However, a number of alternative functions have been proposed. Most of these were intended for specific applications and therefore have not affected the popularity of the power function to any degree. For instance, Carr (1946) proposed an S-type function which was based on the assumption of a gradual start-up. Guibert (1945) proposed a complicated multi-parameter function with several restrictive assumptions.

More recent work holds somewhat more promise. Among these, DeJong (1957) proposed a version of the power function which generates two components, a fixed component which is set equal to the irreducible portion of the task, and a variable component, which is subject to learning. Levy (1965) has presented a new type of firm learning function and shows it to be useful in explaining how firms adapt to new processes, and in isolating the variables that may influence the firm's rate of learning. Levy's learning function reaches a plateau and does not continue to decrease or increase as does the power function. Asher (1956) reports on a variety of different approaches, which were proposed mostly during and immediately following the Second World War. The main drawback of these proposed functions is the difficulty associated with parameter estimation. However, all the proposed alternatives have not been able to dislodge the power function which is apparently still the most common one in use at the present time. It is, of course, possible that the now-common power function formulation is the 'best' algebraic function to describe 'learning' or 'productivity' increases, but on the other hand, this same power function also has some disadvantages.

This paper will propose an alternative algebraic function, an exponential type function, to complement or replace the power function approach. This exponential type function has nearly all the advantages of the power function, but not some of its disadvantages. Hence, it should become a desirable alternative, and may, in addition, generate research interest to develop additional tools and techniques in this area.

To obtain a proper perspective of, and relationship between, the two algebraic functions, the power function will be reviewed first, to be followed by the exponential-type function.

The Power Function

Learning curve literature commonly starts out with the average cost function or more specifically, the cumulative average cost function; this paper shall take a different approach for review purposes, and start out with the total cost function.

$$\mathrm{TC}(x) = \frac{a}{1-b}\,x^{1-b}, \qquad 1$$

where TC(x) is total cost or time incurred up to and including the xth unit,[1] x is the cumulative number of units produced since the start of the process or operation, and a and b are empirically based parameters.

The total cost can be expressed in actual cost figures, or it can be expressed as an index, with the cost of the first unit being, say, equal to one.

If total cost is known, then average cost, or cumulative average cost, can be obtained as follows:[2]

$$\mathrm{AC}(x) = \frac{\mathrm{TC}(x)}{x} = \frac{a}{1-b} x^{-b}, \qquad 2$$

where AC(x) is the average cost of the first x units produced in actual cost figures or measured by the same index used for total cost.

The latter formula is the one commonly known as the 'learning curve' function, although it is directly related to the total cost function, and the marginal or unit cost function to be discussed next.

If we differentiate the total cost function, we obtain the marginal or unit cost function:

$$\mathrm{MC}(x) = \frac{d[\mathrm{TC}(x)]}{dx} = ax^{-b}, \qquad 3$$

where MC(x) is marginal or unit cost per unit for the xth unit[3] in actual cost or in total cost index terms.

1. This paper, from here on, will only mention 'cost' instead of 'cost or time'.

2. The power function used has a negative exponent for marginal and average cost with the result that the cost index declines with increasing productivity. If a positive exponent were used, as some authors do, then the index would rise with increasing productivity, and would be called a productivity or efficiency index.

3. The marginal cost calculated by the mathematical formula is obtained from the slope of the total cost function. Since this total cost function is not linear (does not have a uniform slope) the marginal cost values obtained by the marginal cost function tend to be undervalued, which is especially significant at the start of the process. To compensate for this undervaluation, it is suggested to use the approximation:

$$\overline{\mathrm{MC}}(x) = \frac{\mathrm{MC}(x-1) + \mathrm{MC}(x)}{2} \quad (x = 2, 3, \ldots),$$

or

$$\overline{\mathrm{MC}}(x) = \mathrm{TC}(x) - \mathrm{TC}(x-1) \quad (x = 2, 3, \ldots).$$

For $x = 1$, marginal cost, total cost, and average cost are the same.

The parameter b, in the equations **1**, **2**, and **3** is constrained as follows:

$$0 \leqslant b < 1. \qquad \mathbf{4}$$

If it were not, unrealistic values will be obtained. If $b = 0$, average cost and marginal cost are the same, and total cost will increase at a constant rate. If $b = 1$, then average cost is undefined. There is no constraint on the value of the parameter a; its value depends on the cost index used.

The disadvantage of the above power function formulation can now be illustrated. Analysing the marginal cost and average cost functions, we notice that with increasing x, both marginal and average cost will continue to decrease. This may be a desirable condition for a cost-conscious firm, but it is certainly not realistic, and especially not for marginal cost. In actual practice average cost will continue to decrease after marginal cost becomes a constant but at a very slow rate. In the next section an alternative algebraic formulation, marginal cost will become a constant after a certain number of units have been produced.

The Proposed Exponential Function

The proposed alternative to the power function formulation is an exponential function; it has been derived from the theory of difference equations. If we define the marginal cost of the xth unit, MC(x), to be y_{x-1}, then we can apply the theorem in the next paragraph.[4]

The linear first-order difference equation

$$y_{x+1} = ay_x + b \quad (x = 0, 1, 2, \ldots) \qquad \mathbf{5}$$

(with y_0 given and a and b empirically based parameters) has the unique solution

$$y_x = \alpha a^x + \beta \quad (0 < a < 1) \qquad \mathbf{6}$$

and $\{y_x\}$ converges with limiting value y^*. The parameters α and β are empirically based, and

$$y^* = \frac{b}{1-a}. \qquad \mathbf{7}$$

4. This theorem is a classical one and can be found in most textbooks on difference equations. The author suggests Goldberg (1961, p. 86).

The marginal cost of the first unit, MC(1) = y_0 is assumed known, and will be set equal to the index value one. Hence, from equation **6** we can observe that

$$y_0 = \alpha + \beta = 1{\cdot}00. \qquad \mathbf{8}$$

Furthermore, from equation **6** we can also observe the convergence of $\{y_x\}$. With large x, a^x will approach zero, and

$$y_x = \beta = y^*. \qquad \mathbf{9}$$

To estimate the parameters of equation **6**, we must specify the value of y at convergence, and have one observation of y_x, or estimate one observation of y_x, for

$$y^* < y_x < y_0. \qquad \mathbf{10}$$

In the appendix the parameters for equation **6** are derived.

We can now derive the total cost function from the marginal cost function

$$\text{MC}(x) = \alpha a^{x-1} + \beta \qquad \mathbf{11}$$

by integrating across x as follows,

$$\text{TC}(x) = \int (\alpha a^{x-1} + \beta)\, dx + c$$

$$= \frac{\alpha a^{x-1}}{\ln a} + \beta x + c, \qquad \mathbf{12}$$

where MC(x) is the marginal cost per unit index for the xth unit; TC(x) is the total cost index up to the xth unit in terms of the marginal cost per unit index;[5] α, β, and a are empirically based parameters; and c is a constant to be derived after the parameters have been determined.[6]

Average cost or cumulative average cost can now be derived,

$$\text{AC}(x) = \frac{\text{TC}(x)}{x} = \frac{\alpha a^{x-1}}{x \ln a} + \beta + \frac{c}{x}, \qquad \mathbf{13}$$

5. The total cost values obtained by equation **12** and resultant average cost values tend to be overstated somewhat due to the fact that a continuous function is used. Therefore, it is suggested, when x is small and when accurate total cost values are required, to use the sum of the marginal costs instead of equation **12**.

6. The constant c can be derived by letting MC(1) = TC(1) = AC(1), which provides: $c = \alpha - \alpha/\ln a$.

where AC(x) is the average cost of the first x units produced in marginal cost index terms.

The exponential learning function can be easily converted to a startup or productivity index by substituting $1 - a^{x-1}$ in place of a^{x-1}. If we let P(x) be the productivity function, then

$$P(x) = \alpha(1 - a^{x-1}) + \beta. \quad \textbf{14}$$

Maximum productivity is reached at the index level 1 (one), and the productivity index for the first unit is β.

The exponential functions discussed in this section are not any more difficult to apply than the common and well-known power functions. Although the exponential functions have twice as many parameters as the power functions, these parameters are very easy to calculate. For new processes or new operations, especially those with large amounts of uncertainty, the parameter estimation problem presents the same difficulty with the application of either function. However, the exponential functions have the rational feature of levelling out, to a constant value, for marginal cost. The power function continues to decrease as output increases for both the marginal cost and average cost cases.

A Review of Learning Function Literature

Two of the more recently proposed learning functions have the same levelling-out to a constant value feature as the proposed exponential function. DeJong (1957) proposes a modified version of the power function,

$$MC = a[\beta + (1 - \beta)x^{-b}], \quad \textbf{15}$$

where a and b are parameters, analogous to the power function parameters a and b, and β is the minimum level which marginal cost will approach with large x. Thus β is analogous to the β in the proposed exponential function.

One of the drawbacks of the DeJong function is the difficulty in estimating the parameters a and b. It is assumed that β is known. In the comparison of the various learning functions in the next section, the parameter a will be set equal to 1 (one).

Levy's learning function (1965) presents a still more difficult parameter estimating problem. However, the parameter estimating problems cannot be used as a criticism of Levy's function, since the function is intended to explain how firms adapt to new processes. Levy's function has the form

$$\mathrm{MC} = \left[\frac{1}{\beta} - \left(\frac{1}{\beta} - \frac{x^b}{a}\right) l^{-cx}\right]^{-1}, \qquad 16$$

where a and b are the parameters that are analogous to the power function parameters, c is the third parameter to be estimated, and β is analogous to β in the proposed exponential function, and is assumed known. In order to compare the Levy function with the other functions, the parameters a and b will be set equal to the power function parameters a and b. The function of the parameter c is then to flatten the curve for large values of x.

Two other, although older, learning functions are the ones proposed by Carr (1946) and Guibert (1945). Carr argued that the cumulative average cost curve is best represented by an S-type curve. The initial concave portion of the curve is explained by the common practice, in most production operations, of adding one or several men at a time during the initial startup of a production program. Hence, only relatively small cost savings are obtained during the initial startup with the larger savings coming at a later production stage. Since Carr's learning function applies to a special case we shall not consider it in our analysis.

Guibert introduced the rate of production as a variable affecting unit labor cost. He also viewed the cost curve as having a horizontal asymptote that is approached after a large number of units have been produced. However, the rate of production is also viewed as a determinant of where the decline in unit cost becomes negligible. The assumption, which also underlies our exponential model, that unit cost will eventually follow a horizontal asymptote is general. However, tying it to the rate of production requires a model that becomes restrictive. In addition, Guibert's model centers around the notion that total number of units produced prior to the period in which peak output is reached is approximately equal to the number of units in process during peak production. Because of all the above restrictions and assumptions, the model Guibert has developed is complicated and not very general. It is, therefore, not comparable with the models being discussed in this paper.

The economists have become interested in learning functions because of the effect learning has on production cost functions. It is of interest to point out some of the economists' contributions here as they relate to points usually overlooked by contributors to the traditional learning curve literature. Alchian (1959) agrees with the general proposition that cost of future output declines as total output produced increases. However, he restricts the proposition to apply only to the increase in knowledge and not to a change in technique. The above distinction is usually not made in the industrial and production management literature.

Hirschleifer (1962) responds to some degree to Guibert's model by stating that rate of output will affect cost, but in a different way. He postulates that with volume held constant, marginal cost is a rising function of output. However, with rate of output held constant, marginal cost is a falling function of volume of output. On the latter postulate all of the learning curve literature is based.

Orr (1964) clarified not only Alchian's and Hirschleifer's contributions but also illuminated what has been presented by the traditional learning curve contributors. He starts out by questioning whether any and all production activities can be subsumed in Alchian's model because many firms can modify operations through planning activities which may affect cost considerably. He concludes by stating that Alchian's theory and analysis is of little value in the general case but may hold in industries such as aircraft, job shop, etc., that is, in all those industries where firms produce on order with specified delivery dates. The latter conclusion also applies directly to, and clarifies, all the traditional learning curve literature. Learning curves have not a universal application.

How the Proposed Exponential Function Compares with Other Functions

The proposed exponential function will be compared with the traditional power function and the two modified power function formulations discussed in the previous section. The data used for the comparison consists of eight sets of production data collected by Levy for two offset pressmen on a new offset press. The data is reproduced in Table 1, in modified format. Levy's data showed

Table 1
Production and Cost Data for Two Pressmen on New Press

Pressmen	*Period*	*Cum. output (thousands)*	*Marginal or unit cost index during each period*			
			First shift		*Second shift*	
			With make ready	*Without make ready*	*With make ready*	*Without make ready*
1	1	101	1·000	1·000	1·000	1·000
	2	204	0·951	0·982	0·987	0·925
	3	286	0·924	0·956	0·981	0·861
	4	376	0·928	0·929	0·959	0·815
	5	460	0·898	0·903	0·962	0·812
	6	582	0·871	0·885	0·946	0·817
	7	682	0·874	0·853	0·949	0·810
	8	767	0·835	0·842	0·910	0·806
	9	852	0·781	0·845	0·909	0·792
	10	956	0·776	0·839	0·904	0·772
	11	1046	0·774	0·826	0·897	0·776
	12	1120	0·758	0·821	0·904	0·769
2	1	112	1·000	1·000	1·000	1·000
	2	202	0·980	0·971	0·998	0·973
	3	296	0·939	0·991	0·956	0·935
	4	390	0·906	0·908	0·912	0·862
	5	504	0·909	0·895	0·914	0·878
	6	590	0·897	0·901	0·888	0·848
	7	676	0·893	0·886	0·885	0·848
	8	786	0·886	0·859	0·873	0·817
	9	878	0·886	0·845	0·879	0·806
	10	994	0·862	0·820	0·851	0·809
	11	1090	0·860	0·819	0·845	0·800
	12	1164	0·862	0·801	0·831	0·801

actual productivity for each period, whereas marginal cost data is used for our models.[7] The parameters for all four models were estimated from the marginal cost and cumulative output data

7. In order to use the Levy data the average cost for each period is considered to be the marginal cost at the cumulative output level for that period.

for the first and seventh period. This method of estimating was used because it more nearly approximates actual conditions than parameter estimates based on all observations and derived by the least squares method. Various values of β were used to ensure that an unbiased comparison would result. A typical set of parameters is shown in Table 2.

Table 2

Typical Set of Empirically Based Parameters

Series	β*	*Power model*		*Exponential model*	
		a	*b*	*a*	*b*
1	0·734	1·385	0·071	0·9991	0·0007
2	0·803	1·468	0·083	0·9980	0·0016
3	0·894	1·135	0·027	0·9990	0·0009
4	0·746	1·664	0·110	0·9980	0·0015
5	0·848	1·346	0·063	0·9982	0·0015
6	0·781	1·374	0·067	0·9989	0·0009
7	0·814	1·378	0·068	0·9986	0·0012
8	0·781	1·542	0·092	0·9983	0·0014

Series	*DeJong model*		*Levy model*†		
	a	*b*	*a*	*b*	*c*
1	1·000	0·098	1·385	0·071	0
2	1·000	0·210	1·468	0·083	0
3	1·000	0·101	1·135	0·027	0
4	1·000	0·211	1·664	0·110	0
5	1·000	0·187	1·346	0·063	0
6	1·000	0·113	1·374	0·067	0
7	1·000	0·148	1·378	0·068	0
8	1·000	0·182	1·542	0·092	0

* β is based on a typical γ-value of 0·10, as expressed in the equation: β = cost (period 12) − γ[cost(period 1) − cost (period 12)].

† The value of parameter c was negligible and caused the Levy model to become a replica of the traditional power model.

To obtain a meaningful comparison of the four functions, the sum of squares of deviations (deviations of the actual data from the values based on the algebraic function) for each set of data were calculated. The means and standard deviations of the sums of squares were also calculated and are listed in Table 3 for three different sets of β values.

Conclusions

The derived data supplies some interesting and valuable information. In Table 2 we note that the parameter a for the Exponential Model varies by only a small amount for the eight sets of data. This indicates that the parameter estimation problem may be simplified considerably for similar operations. In addition, the more sensitive parameter b is inversely related to the parameter a in the eight sets of data. This property may be quite valuable if it holds true in all cases. Table 3 indicates that the

Table 3
Measures of Actual Data Deviating from Algebraic Functions

		Sum of squares of deviations	
Function	γ*	*Mean*	*Standard deviation*
Power	—	0·179	0·324
	—	0·179	0·324
	—	0·179	0·324
Exponential	0·02	0·039	0·026
	0·10	0·039	0·026
	0·20	0·039	0·026
DeJong	0·02	0·065	0·032
	0·10	0·064	0·031
	0·20	0·063	0·031
Levy	0·02	0·179	0·324
	0·10	0·179	0·324
	0·20	0·179	0·324

* The constant γ is used to vary β in the equation:

$$\beta = \text{cost (period 12)} - \gamma[\text{cost (period 1)} - \text{cost (period 12)}].$$

proposed exponential function provides a considerably better and consistently better fit (smaller mean and smaller standard deviation) than the three other functions. Only the DeJong function approaches the proposed exponential function. Another important and valuable feature of the proposed exponential function is the fact that varying the value of β has no observable effect on the fit of the algebraic function to the data. The latter feature is important, since β will have to be estimated, and the basis for the estimate will commonly be quite vague.

Appendix

From equation **5**, for $x = 1$,

$$y_1 = ay_0 + b, \tag{A1}$$

for $x = 2$, it follows that

$$y_2 = ay_1 + b = a^2y_0 + b(a + 1) \tag{A2}$$

and for any x.

$$y_x = a^x y_0 + b \sum_{i=0}^{x-1} a^i. \tag{A3}$$

But $y_0 = 1{\cdot}00$, and the geometric progression can be summed, so that we obtain

$$y_x = a^x + b\left(\frac{a^x - 1}{a - 1}\right). \tag{A4}$$

From equations **7** and **9** we can derive

$$b = \beta(1 - a). \tag{A5}$$

From equations **A4** and **A5** we find that

$$a^x = \frac{y_x - \beta}{1 - \beta}, \tag{A6}$$

and

$$a = \text{antilog}\left[\frac{\log(y_x - \beta) - \log(1 - \beta)}{x}\right]. \tag{A7}$$

Suppose that y_x converges at $y^* = \beta = 0{\cdot}20$. Then it follows from equation **8** that $a = 0{\cdot}80$. Also suppose that $x = 60$, $y_x = 0{\cdot}60$. Then from equation **A6** we find that $a = 0{\cdot}9935$.

To determine the value of y_x at a given value of x we use the total cost incurred up to the $(x + 1)$th unit processed or produced and subtract from it the total cost incurred up to the xth unit.

References

ALCHIAN, A. (1959), 'Costs and outputs', in M. Abramoritz *et al.*, *Economic Resources*, Stanford University Press.

ASHER, H. (1956), *Cost–Quantity Relationship in the Airframe Industry*, RAND Corporation.

BALOFF, N. (1967), 'Estimating the parameters of the startup model – an empirical approach', *J. Industr. Eng.*, vol. 18, no. 4, pp. 248–53.

CARR, G. W. (1946), 'Peacetime cost estimating requires new learning curves', *Aviation*, vol. 45, pp. 76–7.

DEJONG, J. R. (1957), 'The effects of increasing skills on cycle time and its consequences for time standards', *Ergonomics*, vol. 1, no. 1, pp. 51–9.

GOLDBERG, S. (1961), *Introduction to Difference Equations*, Wiley.

GUIBERT, P. (1945), *Le Plan de Fabrication Aeronautique*, Dunod. Translated as *Mathematical Studies of Aircraft Construction*, by Central Air Documents Office, Wright-Patterson Air Force Base, Ohio.

HIRSCHLEIFER, J. (1962), 'The firm's cost function: a successful reconstruction', *Journal of Business*, vol. 35, pp. 235–55.

LEVY, F. K. (1965), 'Adaption in the production process', *Management Science*, vol. 11, no. 6, pp. 136–54.

ORR, D. (1964), 'Costs and outputs: an appraisal of dynamic aspects', *Journal of Business*, vol. 37, pp. 51–60.

WRIGHT, T. P. (1936), 'Factors affecting the cost of airplanes', *Journal of the Aeronautical Sciences*, vol. 3, pp. 122–8.

12 L. E. Freund and T. L. Sadosky

Linear Programming Applied to Optimization of Instrument Panel and Workplace Layout

L. E. Freund and T. L. Sadosky, 'Linear programming applied to optimization of instrument panel and workplace layout', *Human Factors*, vol. 9 (1967), pp. 295–300.

The layout of a workplace or instrument panel should provide for optimal operator performance. The literature shows a tendency to formulate general human factors principles for workplace arrangement and to offer them as recommendations to the design engineer. For example, the proper arrangement of displays and controls is a basic problem in workplace layout. Among the principles cited for the design engineer are: (1) placement of the most important displays and controls in the optimum working and viewing area and (2) organization on the basis of function, frequency of use, or sequence of operation. Effective application of such principles requires that human factors criteria be expressed in a quantitative form useful to the design engineer.

The purpose of this paper is to demonstrate the applicability of several linear programming algorithms to the general problem described above. The problem will be formulated in two different ways: (1) optimization of layout based on minimum 'utility cost' and (2) optimization of layout based on minimum eye travel. Example problems will be presented to illustrate the benefits or difficulties of each approach. The specific algorithms used for solution may be found in several linear programming texts, such as Dantzig (1963).

Optimization Based on Minimum 'Utility Cost'

Work place layout

Consider the problem of optimizing the arrangement of a workplace, for example an airplane cockpit. Fitts (1947), investigating the problem of pilot errors in aircraft control manipulation, obtained data relating to the desirability of various cockpit areas for the placement of controls. In effect, he subdivided the cockpit into eight distinct areas as shown in Figure 1. His data reflected the relative accuracy of the pilot in a blind reach to each of the eight areas. Subsequently, Deininger (1958) collected data on the frequency with which the pilot made certain task

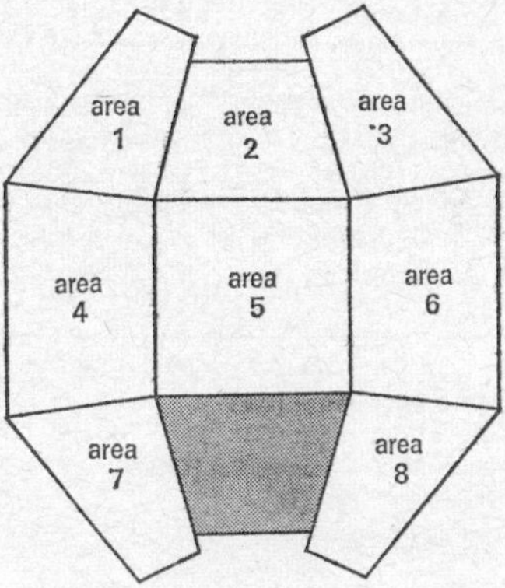

Figure 1 Workplace areas (Fitts, 1947)

responses in flying simulated cargo missions. Combining these two reports, it is possible to construct a 'utility cost' rating for placing each of the eight controls in each of the eight areas where 'utility cost' is defined as the product of the frequency of use and the area mean accuracy score. (See Table 1.) For example, the cost of placing control M in area 1 is 3·49 × 13 or 45·37, and the cost of placing control E in area 4 is 3·13 × 30 or 93·90.

The problem is now in a form which can be treated by an assignment algorithm where the objective is to assign one control to each area at minimum total utility cost. While solution of this problem is possible by the Hungarian method, the special structure of the problem (each entry in the tableau is a product of frequency times mean accuracy score) indicates that the product method is applicable (Huebner and Ryack, 1961). In this method, the controls are arranged in order of decreasing frequencies and the areas are arranged in order of increasing mean error scores. Table 1 has been arranged in this manner. The optimal solution,

Table 1

Utility Cost Tableau (Fitts, 1947, and Deininger, 1958)

		Controls†							
Mean accuracy score	*Area*	*AP*	*I*	*E*	*T*	*C*	*AC*	*M*	*G*
		*115***	40	30	27	20	15	13	12
2·14	5	246·10*	85·60	64·20	57·78	42·80	32·10	27·82	25·68
2·46	2	282·90	98·40*	73·80	66·42	49·20	36·90	31·98	29·52
2·76	8	317·40	110·40	82·80*	74·52	55·20	41·40	35·88	33·12
3·03	7	348·45	121·20	90·90	81·81*	60·60	45·45	39·39	36·36
3·07	6	353·05	122·80	92·10	82·89	61·40*	46·05	39·91	36·84
3·13	4	359·95	125·20	93·90	84·51	62·60	46·95*	40·69	37·56
3·21	3	369·15	128·40	96·30	86·67	64·20	48·15	41·73*	38·52
3·49	1	401·35	139·60	104·70	94·23	69·80	52·35	45·37	41·88*

* Optimal solutions
** Frequency
† Abbreviations:
Mixture (M)
Throttle (T)
Elevator Trim (E)
Auto Pilot Panel (AP)
Cross Pointer Set (C)
Gyro-Mag. Compass set (G)
Intercom Panel (I)
Auto. Compass panel (AC)

found by making assignments on the principle diagonal of the tableau, is shown in the Table by starred (*) numbers. A proof of this method is given by Huebner and Ryack (1961).

Relying upon hypothetical data, we next consider possible extensions of this problem as follows. Suppose we have c controls

Table 2

Initial Tableau for Transportation Algorithm

		Areas	*1*	*2*	*3*
		Capacity	*1*	*2*	*1*
		Mean error score	*5*	*2*	*4*
Controls	*Frequency of use*				
a	1		5	2	4
b	2		10	4	8
c	3		15	6	12
d	4		20	8	16

to assign to n areas, where $n < c$, where each control occupies equal space, but where the areas are of differing size. As an example, let $n = 3$ and suppose that there are capacities restricting the number of instruments that can be placed in a given area as indicated in Table 2. We desire to place a total of 4 controls in the 3 areas. Table 2 shows the frequency of use associated with each control. The cell entries in Table 2 represent the product of frequency of use and mean accuracy score. The problem is now in a form which can be solved as a transportation problem. The solution is shown in Table 3, where an optimal assignment is indicated by starred entries. Control *a* is assigned to area 1, control *b* to area 3, and controls *c* and *d* to area 2.

Table 3
Final Tableau for Transportation Algorithm

		Areas	*1*	*2*	*3*
		Capacity	*1*	*2*	*1*
		Mean error score	5	2	4
Controls	*Frequency of use*				
a	1		0*	5	1
b	2		0	2	0*
c	3		1	0*	0
d	4		4	0*	2

* Optimal assignments.

As a further extension, we allow the number of each type of control to be assigned to vary, and add additional capacity restrictions relating to placing certain controls in certain areas, as shown in Table 4. The area capacities and mean accuracy scores are also shown in Table 4. We can now formulate the problem as a capacitated transportation problem, where again the cell entries in the matrix represent the cost of assigning control i to area j. For example, the problem is constrained so as to allow placement of no more than two type *a* controls to areas 1, 2, or 3 in the optimal solution; these restrictions are represented by

Table 4

Initial Tableau for Capacitated Transportation Algorithm

			Area		
			1	*2*	*3*
			Capacity		
			4	*2*	*3*
			Mean error score		
			5	*2*	*4*
Controls	*Frequency of use*	*Number to be assigned*			
a	1	4	$^{2}5$	$^{2}2$	$^{2}4$
b	2	3	10	4	8
c	3	2	15	6	12

Table 5

Final Tableau for Capacitated Transportation Algorithm

			Area		
			1	*2*	*3*
			Capacity		
			4	*2*	*3*
			Mean error score		
			5	*2*	*4*
Controls	*Frequency of use*	*Number to be assigned*			
a	1	4	$^{2}1^{2}$	$^{2}4$	$^{2}2^{2}$
b	2	3	0^{2}	0	0^{1}
c	3	2	3	0^{2}	2

min. cost = 58

numbers in the upper left corner of the appropriate cells. The final tableau is shown in Table 5 and the optimal solution is to assign two type *a* controls to area 1, two type *a* controls to area 3, two type *b* controls to area 1, one type *b* controls to area 3, and two type *c* controls to area 2. We assume that the frequency of use of each of the controls of the *i*th type is equal (if not, we would subdivide the *i*th type of control into equivalent frequency subclasses).

Instrument arrangement

The problem of the optimal arrangement of n instruments in a fixed pattern on a control panel can also be treated as an assignment problem. In the discussion below, we assume that the panel positions available for instrument assignment are known and the instruments to be assigned are of equal size. We also continue to work with hypothetical data.

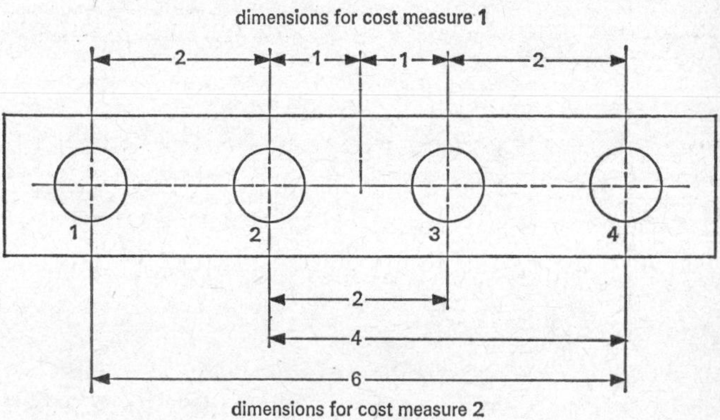

Figure 2 Linear arrangement with distance measured between positions and from panel center

Consider the problem of a straight line arrangement of four instruments on a control panel. The panel could be represented as in Figure 2. Using the 'utility cost' concept, we could measure the distance component of cost of each position in two ways (see Figure 2): (1) as a function of the distance, in inches, from the geometric center of the panel to the center of each instrument (where the panel center is assumed to be the point of normal eye fixation). This approach, in effect, implies a larger probability of instrument misinterpretation as the distance from the panel center increases; and (2) as a function of the sum of the distances from a particular position to every other position on the panel. In both cases the distance component of cost will be termed the 'distance value'.

Table 6
Initial Tableau for Assignment Algorithm

		Position number			
		1	*2*	*3*	*4*
		Distance value			
		3	*1*	*1*	*3*
Instrument no.	*Frequency of use/hr.*				
1	25	75	25	25	75
2	13	39	13	13	39
3	16	48	16	16	48
4	19	57	19	19	57

Approach using cost measure 1. Using the panel center to instrument center as the distance value, it is additionally necessary to have data describing the frequency of use of each instrument. Assume the data is as shown in Table 6. The assignment matrix is also given in Table 6, where the cell entries in the matrix represent the cost of placing instrument i in position j (i.e. distance value (from Figure 2) times the frequency of use per hour). Using the Hungarian method of solution, we can derive the final tableau given in Table 7. The optimal assignment is given by the starred entries in the table. The minimum utility cost is then

$$48 + 25 + 19 + 39 = 131.$$

The optimum assignment is Instrument 1 to position 2, Instrument 2 to position 4, Instrument 3 to position 1, and Instrument 4 to position 3.

This solution could also be obtained using the product method. First we would arrange the instruments in descending order according to decreasing frequency of use. We would next arrange the positions in order of increasing distance value. We would then obtain a solution comparable to that shown in Table 7 by making assignments along the principle diagonal.

Table 7
Final Tableau for Assignment Algorithm

		Position number			
		1	*2*	*3*	*4*
		Distance measure			
		3	*1*	*1*	*3*
Instrument number	*Frequency of use/hr.*				
1	25	18	0*	0	18
2	13	0	6	6	0*
3	16	0*	0	0	0
4	19	6	0	0*	6

* Optimal assignments.

Approach using cost measure 2. Here, the distance value of any position is the sum of the distances from that position to every other position on the panel. For example, in Figure 2 the distance value of positions 1 and 4 would be $2+4+6=12$ inches each, and the distance value of positions 2 and 3 would be $2+4+2=8$ inches each.

With the structure, it is possible to consider the probability of transition between instruments a and b as a contributing factor to the optimal arrangement. This consideration is not possible under the approach using cost measure *1* above. The example presented will utilize the probability of transition matrix shown in Table 8. Structured as an assignment problem, the initial tableau would appear as in Table 9, where the entry in the (i, j)th cell of the matrix represents the distance value of the posi-

Table 8
Transition Probability Matrix

From instrument number	*To instrument number 1*	*2*	*3*	*4*	*Transition probability sub-total*	*Away from panel*	*Total*
1	0	0·1	0	0·3	0·4	0·6	1·0
2	0·2	0	0·6	0·1	0·9	0·1	1·0
3	0·4	0·5	0	0·1	1·0	0	1·0
4	0·2	0·3	0·4	0	0·9	0·1	1·0

tion j times the probability of transition from instrument i to any other instrument in the panel. For example, this probability is shown in Table 8 as the transition probability subtotal. Placing

Table 9
Initial Tableau for Assignment Algorithm

		Position number			
		1	*2*	*3*	*4*
		Distance factor			
		12	*8*	*8*	*12*
Instrument number	*Transition probability sub-total*				
1	0·4	4·8	3·2	3·2	4·8
2	0·9	10·8	7·2	7·2	10·8
3	1·0	12·0	8·0	8·0	12·0
4	0·9	10·8	7·2	7·2	10·8

instrument 1 in position 1 would have an associated cost of $0{\cdot}4 \times 12 = 4{\cdot}8$. This value appears in the (1, 1) cell in the matrix in Table 9. Again, using the Hungarian method of solution, we can derive a final tableau which is presented in Table 10. The minimum utility cost is then $4{\cdot}8 + 8 + 7{\cdot}2 + 10{\cdot}8 = 30{\cdot}8$. An optimum assignment is Instrument 1 to position 1, Instrument 2 to position 3, Instrument 3 to position 2, and Instrument 4 to position 4.

Table 10
Final Tableau for Assignment Algorithm

		Position number			
		1	*2*	*3*	*4*
		Distance factor			
		12	*8*	*8*	*12*
Instrument number	*Transition probability sub-total*				
1	0·4	0*	2·0	2·0	0
2	0·9	0	0	0*	0
3	1·0	0·4	0*	0	0·4
4	0·9	0	0	0	0*

* Optimal assignments.

Summary of utility cost method

From the above examples, it is apparent that when a utility cost method is used, the problem of instrument panel design or workplace layout can be reduced to a transportation or assignment problem which can be solved by standard methods. While rather simple designs have been used in the above examples, the methods described can easily be extended to treat more realistic situations.

Optimization based on Minimum Eye Travel Distance

It is noted that an optimal panel design might be one which minimizes the operator's eye travel in using the panel. The objective function will take the form:

$$\text{Min } z = \sum x_{ij} p_{ij},$$

where x_{ij} is the distance from instrument i to instrument j and p_{ij} is the probability of transition from instrument i to instrument *j*. This problem was formulated for simplex solution. The panel design considered was a square four-instrument arrangement shown in Figure 3. The probability of transition matrix is given in Table 8. We wish to:

$$\text{Min } z = 0{\cdot}1x_{12} + 0{\cdot}3x_{14} + 0{\cdot}2x_{21} + 0{\cdot}6x_{23} + 0{\cdot}1x_{24} + 0{\cdot}4x_{31} + \\ + 0{\cdot}5x_{32} + 0{\cdot}1x_{34} + 0{\cdot}2x_{41} + 0{\cdot}3x_{42} + 0{\cdot}4x_{43},$$

which, with the recognition that $x_{ij} = x_{ji}$, can be reduced to

$$\text{Min } z = 0{\cdot}3x_{12} + 0{\cdot}4x_{13} + 0{\cdot}5x_{14} + 1{\cdot}1x_{23} + 0{\cdot}4x_{24} + 0{\cdot}5x_{34}$$

subject to the following restrictions:

$$x_{ij} \geqslant 2 \qquad i < j$$

$$\qquad i = 1, 2, 3$$

$$x_{ij} \leqslant 2{\cdot}8284 \qquad j = 2, 3, 4$$

$$x_{12} + x_{13} + x_{14} + x_{23} + x_{24} + x_{34} = 4(2) + 2(2{\cdot}8284) = 13{\cdot}6568$$

$$\begin{aligned} x_{12} + x_{13} &\leqslant 4{\cdot}8284 \\ x_{13} + x_{14} &\leqslant 4{\cdot}8284 \\ x_{12} + x_{14} &\leqslant 4{\cdot}8284 \\ x_{23} + x_{24} &\leqslant 4{\cdot}8284 \\ x_{24} + x_{34} &\leqslant 4{\cdot}8284 \\ x_{12} + x_{23} &\leqslant 4{\cdot}8284 \\ x_{12} + x_{24} &\leqslant 4{\cdot}8284 \\ x_{13} + x_{23} &\leqslant 4{\cdot}8284 \\ x_{13} + x_{34} &\leqslant 4{\cdot}8284 \\ x_{14} + x_{24} &\leqslant 4{\cdot}8284 \\ x_{14} + x_{34} &\leqslant 4{\cdot}8284 \\ x_{23} + x_{34} &\leqslant 4{\cdot}8284. \end{aligned}$$

The first pair of restrictions set upper and lower bounds on values of x_{ij} to maintain feasibility. The third restriction also is included to maintain feasibility of the solution by restricting the sum of the possible distances. In this problem, we are essentially asking which instruments should be 'caddy-corner' from each other. The remaining feasibility problem is to define which instruments can be across the diagonals from each other in a feasible solution. This is done by noting that the sum of pairs of distances which have a common digit in their subscript must be either 4·0 (both distances are along the rim), or 4·8284 (one distance is along the rim and the other distance is across a diagonal). These restrictions are enumerated in the last group of constraints given above.

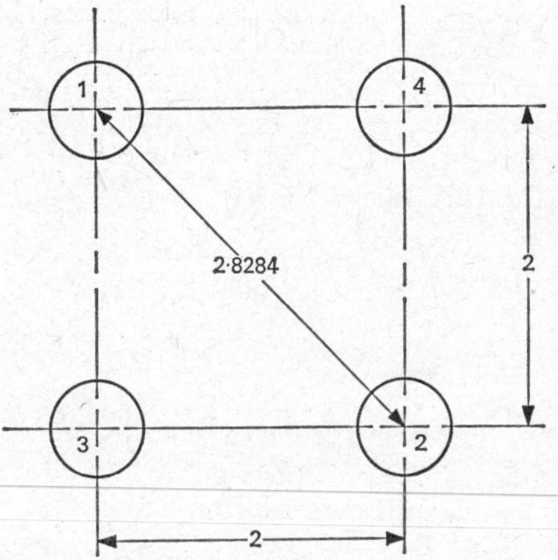

Figure 3 Square arrangement of four instruments

Even though the number of instruments was small, the problem demanded constraints which made it large for hand computation. Thus, a computer programmed simplex routine was utilized, and the results are as follows:

$$x_{12} = 2{\cdot}8284, \quad x_{13} = 2{\cdot}0, \quad x_{14} = 2{\cdot}0, \quad x_{23} = 2{\cdot}0,$$
$$x_{24} = 2{\cdot}0, \quad x_{34} = 2{\cdot}8284, \quad z_{opt} = 7{\cdot}2627.$$

Knowing the optimal feasible distances between instruments, we can then construct a suitable panel. For this example problem the optimal arrangement of the four instruments is shown in Figure 3 by the numbers 1 through 4. In our investigation of the problem of arranging 4 instruments in a linear fashion (as shown in Figure 2) utilizing the simplex algorithm, we were unable to establish any systematic procedure for specifying constraint inequalities for a feasible solution.

Summary

The following statements seem justified as a result of this analysis:

1. The workplace layout or control panel optimization problem can be approached using the utility cost method described. It has been demonstrated that this approach will lend itself to many problems of this type enabling optimization by the product method, the Hungarian method, or the assignment or capacitated transportation algorithms, where the algorithm applied depends upon the problem structure.

2. The objective of minimization of total eye travel distance requires a complex system of constraint equations which appear to limit its usefulness as a practical technique.

References

DANTZIG, G. B. (1963), *Linear Programming and Extensions*, Princeton University Press.

DEININGER, R. L. (1958), 'Process sampling, workplace arrangements and operator activity levels', USAF: WADC., Engineering Psychology Branch, Unpublished MS.

FITTS, P. M. (1947), 'A study of location discrimination ability', in P. M. Fitts (ed.), *Psychological Research on Equipment Design*, U.S. Government Printing Office, Army Air Force Aviation Psychology Program Research Reports no. 19.

HUEBNER, W. J., and RYACK, B. C. (1961), 'Linear programming and workplace arrangement, solution of assignment problems by the product technique', USAF: WADC, TR 61–143.

Part Three **Operations Management**

Technological and methodological inventions lead to new operations, different policies, higher-level control systems and different descriptions of the production manager's job. Consider the articles in this section.

Before 1950 production managers hadn't even heard of the operations management concepts dealt with in this section. Today they are widely known and practised. Since they interrelate with process decision, the total face of the production system is changing.

Terms like information theory, decision trees, queueing theory, human engineering and learning theory are some examples of the name-dropping game in process design and operations. We can show generality of these new methods, even for old process problems. But, textbook techniques often float about without any hope of anchoring in the student's mind. Insight usually develops when the reader comes face to face with the applicability of what he is reading. The articles in this section were chosen to reflect operations technique in application. They show that one need not be concerned with technique for the sake of technique after it has been sufficiently well-invented and explored.

13 J. W. Gavett

Planning Models

Excerpt from J. W. Gavett, *Production and Operations Management*, Harcourt, Brace & World, 1968, pp. 454–74.

Planning Models

Leveled production

Among the many possible alternative production plans, there are two extreme possibilities. The first is to meet the fluctuations in demand directly with corresponding changes in the employment level. This includes the hiring and layoff of single shift personnel, or the addition or elimination of extra shifts. Such a policy is mandatory in some seasonal businesses dealing with products or services that cannot be backlogged or inventoried. Otherwise, while such a policy is possible, it has been noted that it can be very costly. Besides the direct and indirect costs previously cited, there are the social costs, which, while obvious, are not easy to measure.

The second possibility is to stabilize the labor force at some predetermined level and meet changes in demand mainly by altering inventories, backlogs, overtime, and other means. This is called leveled production.

In many cases, a production plan may consist of some mixture of these two extremes. A basic policy of leveled production can be augmented by temporary periods of overtime and added shifts or part-time personnel. Periods of low demand can be utilized for internal service or maintenance operations or complemented with vacation scheduling, besides reducing backlogs or increasing inventories.

Linear decision rules for production planning

Perhaps the most comprehensive experiment in production planning was conducted at the Carnegie Institute of Technology in 1955 (Holt *et al.*, 1960). The problem was to determine a decision rule for making production and labor-force decisions in successive time periods that would minimize the expected value of total costs over a large number of periods. The study resulted in the formulation of a mathematical model consisting of a set of linear decision rules. We will consider the basic features of the method.

Studies of cost behavior. The method employs a rigorous analysis of the behavior of the relevant planning costs with changes in the planning variables. The following notation is used:

W_t = the number of workers required for the period t; $t = 1, 2, \ldots, T$

P_t = the number of units to be produced in period t

I_t = the inventory minus backlogs at the end of period t

S_t = forecast of the number of units to be ordered for shipment during the period t

C_k = cost coefficients to be determined for a given plant or situation.

Figure 1 shows the hypothetical behavior of the planning costs (solid line) and a quadratic approximation to these costs in each case. Each cost function expresses the change in cost corresponding to a change in the planning variable from one month to the next. The values of the cost coefficients must be determined individually for each plant in which the model is used.

From such analysis, a total cost function of the following kind is derived:

$$C_T = \sum_1^T C_t$$

and

$C_t = [(C_1 W_t) +$	Regular payroll costs
$+ C_2 (W_t - W_{t-1})^2 +$	Hiring and layoff costs
$+ C_3 (P_t - C_4 W_t)^2 + C_5 P_t - C_6 W_t +$	Overtime costs
$+ C_7 (I_t - C_8 - C_9 S_t)^2]$	Inventory connected costs

subject to constraints

$$I_{t-1} + P_t - S_t = I_t.$$

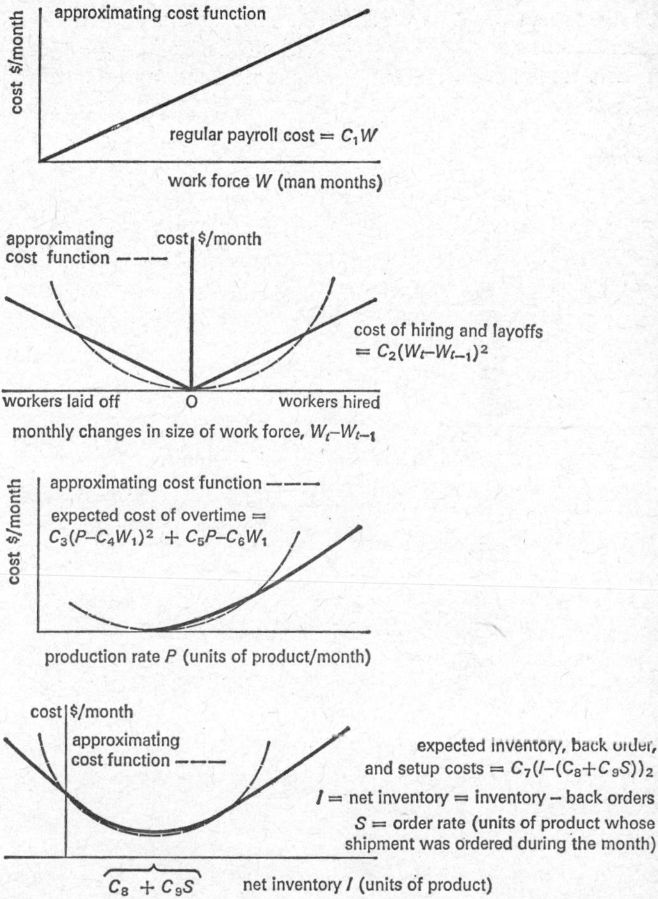

Figure 1 Hypothetical cost behavior as a function of planning variables. From Holt *et al.* (1955)

Once such a cost function is obtained, the decision rule that minimizes the expected value of C_T is obtained, by differentiating with respect to each decision variable. This results in a set of *linear* equations whose solution, in turn, yields the decision rule. Note that the assumption of quadratic cost functions results in linear equations after differentiating, and finally in linear decision rules.

The resulting linear decision rules are of the following form:

$$P_t = \sum_{t=1}^{T} \lambda_t S_t + k_1 W_{t-1} + k_2 - k_3 I_{t-1} \qquad \mathbf{1}$$

$$W_t = k_4 W_{t-1} + k_5 - k_6 I_{t-1} + \sum_{t=1}^{T} a_t S_t. \qquad \mathbf{2}$$

Consider equation **1** for calculating the production level P_t. First, it is a function of the forecasted sales for T periods in the future. λ_t is a specific weight given to each forecast. The more immediate period forecasts are given a higher weight than the more distant ones. The implication is that it is not economical to produce currently for shipments that are far into the future, because of inventory costs. Second, production is also a direct function of the employment level at the beginning of the period, thus reflecting the desirability of avoiding the cost of changing the level. Finally, if the inventory at the beginning of the period is high, then production may be reduced in order to lower the inventory.

With respect to equation **2** for the employment level W_t, we note that it is also a function of the forecasted sales for which a weight a_t is used. Typically, the value of a_t declines as t increases. However, the weights extend further into the future before they become negligible, such that forecasts into the more distant future influence employment level decisions. For example, in one study of a paint factory, the values of λ_t and a_t were

t	1	2	3	4	5	6
λ_t	+0·458	+0·233	+0·111	+0·046	+0·014	−0·001
a_t	+0·010	+0·009	+0·007	+0·005	+0·004	+0·003

t	7	8	9	10	11	12
λ_t	−0·007	−0·008	−0·008	−0·007	−0·005	−0·004
a_t	+0·002	+0·002	+0·001	+0·001	+0·001	$+0{\cdot}000^{+}$

Thus, changes in sales that will take place in the distant future are reflected in present changes in employment levels, whereas more immediate changes in sales are reflected in current production level changes through overtime.

The two equations yield a dynamic response system to changes in demand. The production level for one period influences the inventory level, which in turn affects the employment level for the next period, which then affects the production level for the third period. Short-term fluctuations in demand, such as random variations, are absorbed by varying production levels, whereas long-term trend and cyclical changes are met by gradual changes in employment levels.

This method of planning production was tested in several actual situations by the originators of the method. One study was made of a paint factory in which the decision rules were applied to a history of sales over the 1952 to 1954 period. A moving-average forecast was used to provide the forecast requirements for the model. If the rules had been used during the study period, the reduction in cost, as compared to the company's actual performance, would have been 7·8 per cent, as shown in Table 1.

Table 1

Comparison of Performance using Decision Rules with Actual Performance *

Costs (thousands of dollars)		*Actual company performance*	*Decision rule (moving-average forecast)*
Regular payroll		$1256	$1149
Overtime		82	95
Inventory		273	298
Back orders		326	246
Hiring and layoffs		16	12
Total cost	$	1953	1800
	%	108·5	100

* From Holt *et al.* (1955).

This method of using linear decision rules has not been widely employed. Undoubtedly, this is due to the amount of analysis required to obtain the necessary cost data, as well as to the mathematical sophistication of the method. However, the derivation of the linear rules can be accomplished on a computer with reasonable efficiency. This method represents a very thorough approach to an important problem and should not be ignored because of its mathematical complexity.

Linear programing

A number of linear programing approaches have been taken to solve the problem. A rather straightforward use of the simplex method of a single facility and single product is the following (Magee and Boodman, 1967, pp. 369–73). Let

i = index of planning period; $i = 1, 2, \ldots, k$
$d(i)$ = forecasted demand in units for the ith period
$h(i)$ = maximum number of units that can be produced on regular time in ith period prior to change in employment level
$w(i)$ = maximum number of units that can be produced on overtime in ith period
$r = w(i)/h(i)$
$p(i)$ = number of units produced on regular time in ith period
$y(i)$ = number of units produced on overtime in ith period
b = overtime marginal cost per unit of product
c = inventory carrying charge per unit per period
a' = cost to add one product unit of capacity to production
a'' = cost to remove one product unit of capacity from production (note that both a' and a'' are equivalent to employment level changes)
$u(i)$ = number of units of capacity added in period i
$v(i)$ = number of units of capacity removed in period i.

Note that all capacity and cost measures have been expressed in terms of units of product. This permits an expansion of the method for the case of more than a single product.

We now have the following restrictions:

1. Restrictions on regular production

$$p(1) - u(1) + v(1) \leqslant h(1)$$
$$p(2) - u(1) - u(2) + v(1) + v(2) \leqslant h(2)$$
$$\vdots \qquad \vdots \qquad \vdots \qquad \vdots$$
$$p(k) - u(1) - \cdots - u(k) + v(1) + \ldots + v(k) \leqslant h(k).$$

2. Restriction on overtime production

$$\begin{aligned}
&y(1) + r[-u(1) + v(1)] \leqslant w(1) \\
&y(2) + r[-u(1) - u(2) + v(1) + v(2)] \leqslant w(2) \\
&\qquad\vdots \\
&y(k) + r[-u(1) - \cdots - u(k) + v(1) + \cdots + v(k)] \leqslant w(k).
\end{aligned}$$

3. Restrictions on inventories (assuming no opening inventory in first period)

$$\begin{aligned}
p(1) + y(1) &\geqslant d(1) \\
p(1) + p(2) + y(1) + y(2) &\geqslant d(1) + d(2) \\
&\vdots \\
\sum_{j=1}^{i} p(j) + \sum_{j=1}^{i} y(j) &\geqslant \sum_{j=1}^{i} d_j \quad (i = 1, 2, \ldots, k).
\end{aligned}$$

4. Capacity added and removed

$$\begin{aligned}
u(1) - p(1) &\geqslant - p(0) \\
u(2) - p(2) + p(1) &\geqslant 0 \\
&\vdots \\
u(k) - p(k) + p(k-1) &\geqslant 0
\end{aligned}$$

and similarly, $\quad v(k) + p(k) - p(k-1) \geqslant 0.$

Now the objective function to be minimized is

$$C(p, y, u, v) = a' \sum_{1}^{k} u(i) + a'' \sum_{1}^{k} v(i) + \alpha(p, y)$$

where

$$\begin{aligned}
\alpha(p, y) &= b \sum_{1}^{k} y(i) + c\left[\sum_{i=1} \sum_{j=1}^{i} p(j) + y(j) - d(j) \right] \\
&= \sum_{1}^{k} cp(i)(k - i + 1) + \sum_{1}^{k} y(i)[b + c(k - i + 1)] + D
\end{aligned}$$

and $\quad D = - c \sum_{i=1}^{k} \sum_{j=1}^{i} d(j).$

Dynamic programing

Another approach to production planning uses dynamic programing. For our example we will use the following notation. Let

P_t = the number of regular production hours used in period t

I_t = the inventory level, in equivalent hours, at the beginning of period t

D_t = the period t demand for the product in equivalent hours

ϕ_t = the number of hours of overtime used in period t

S_t = an index of changes in production level (employment level)

$$S_t = \begin{cases} 1 \text{ if the production level is changed from period } t-1 \text{ to period } t \\ 0 \text{ if no change in production level takes place} \end{cases}$$

$c_t(P)$ = the cost per hour of regular production in period t

$C_t(I)$ = the cost of carrying one hour of inventory for period t.

We will assume that the inventory at the beginning of the period is carried for the full period.

$c_t(\phi)$ = the cost per hour of overtime in period t

$c_t(E)$ = the cost of shifting the regular production level from period $t-1$ to period t. We will assume that the production level in any period can assume one of two values.

$C_t(1)$ = the total cost of regular production in period t

$= P_t \, c_t(P)$

$C_t(2)$ = the total inventory cost in period t

$= I_t \, c_t(I)$

$C_t(3)$ = the total cost of overtime in period t

$= \phi_t \, c_t(\phi)$

$C_t(4)$ = the total cost of changing production level in period t

$= S_t\, c_t(E)$.

$TC_t(P_t, \phi_t: P_{t-1}, I_t)$ = the total cost for period t given the choice of the regular production level P_t and overtime ϕ_t conditioned by a production level change of $P_{t-1} - P_t$ and by a beginning period inventory of I_t

$= C_t(1) + C_t(2) + C_t(3) + C_t(4) +$ minimum $TC_{t+1}(: P_t, I_{t+1})$.

Now consider an example with arbitrarily restrictive assumptions. Let there be four planning periods. Regular production levels P_t can be changed by going from one shift (1000 hours) to two shifts (2000 hours) or vice versa. The present employment level (preceding period 1) is 2000 hours. Overtime for any period ϕ_t can have the value zero or 500 hours. The beginning period inventory I_t can assume the alternative values of 500, 1000, or 1500 hours. The inventory at the end of period 4 must be 500. The inventory at the beginning of period 1 is zero. The following data are relevant:

		$c_t(P)$				
t	D_t	*One shift*	*Two shifts*	$c_t(\phi)$	$c_t(I)$	$c_t(E)$
1	1000	\$2·50	\$3·00	\$4·00	\$1·00	\$5000
2	2500	2·50	3·00	4·00	1·00	5000
3	2000	2·50	3·00	4·00	1·00	6000
4	1500	3·00	4·00	5·00	1·00	6000

The problem is to determine the regular production hours (number of shifts) and overtime for each of four periods, which results in the minimum total cost of production, overtime, inventory, and production level changes. That is, we wish to minimize $TC(1)$, subject to

$$P_0 = 2000$$
$$I_1 = 0$$
$$I_5 = 500$$
$$I_{t+1} = I_t + P_t + \phi_t - D_t.$$

Table 2
Chart for Period 4 Decision

			P_4 1000	1000	2000	2000
			ϕ_4 0	500	0	500
P_3	I_4					
1000	500	$C(1)$	—	3·0	—	—
		$C(2)$		0·5		
		$C(3)$		2·5		
		$C(4)$		0·0		
		$TC(t+1)$		0·0		
				6·0		
2000	500		—	3·0	—	—
				0·5		
				2·5		
				6·0		
				0·0		
				12·0		
1000	1000		3·0	—	—	—
			1·0	—	—	—
			0·0			
			0·0			
			0·0			
			4·0			
2000	1000		3·0	—	—	—
			1·0			
			0·0			
			6·0			
			0·0			
			10·0			
1000	1500		—	—	—	—
2000	1500		—	—	—	—

For any period, the decision alternatives are to use one shift only, one shift with overtime, two shifts only, or two shifts with overtime. We commence with period 4. The four alternatives are listed across the top of Table 2. The rows of the table are labeled with possible combinations of period 3's employment level and period 4's opening inventory. Each cell in the table indicates the total cost $TC(4)$ of the production plan labeling the column. For example, if, in period 4, we produce at a level of 1000 hours with 500 hours of overtime, the total cost will be \$3000 production, \$500 inventory, \$2500 overtime, and zero for a change in employment level, since we assumed 1000 hours in period 3. Since the ending inventory for the fourth period is specified as 500 hours and the demand for that period is 1500 hours, certain plans are inadmissible, as indicted by the dashes.

The chart for period 3 is shown in Table 3. Note that each element is standardized as follows:

	$\lvert I_{t+1}$
$C_t(1)$	
$C_t(2)$	
$C_t(3)$	
$C_t(4)$	
min. $TC(t+1)$	

For example, if we produce at 2000 hours without overtime, and assuming the opening inventory is 500 and the period 2 production level was 1000, then the total cost would be production \$6000, inventory \$500, overtime zero, and change in production level \$6000. Since this plan would result in an ending inventory of 500 hours, we must add to this the minimum fourth period plan for the conditions $P_3 = 2000$ and $I_4 = 500$, which is \$12,000. Thus, the total cost is \$24,500. Note that we carry back each period the optimum costs from the following period for the given value of P_t and I_{t+1}. The student may continue the problem until a production plan is derived.

Applications. In reviewing the problem of production planning under conditions of seasonal demand, Vergin conducted a study of eight firms having highly seasonal products (Vergin, 1966, pp. 260–66). He noted that in most of the firms, management

Table 3
Chart for Period 3 Decision

<table>
<tr><td></td><td>P_3</td><td>1000</td><td>1000</td><td>2000</td><td>2000</td></tr>
<tr><td></td><td>ϕ_3</td><td>0</td><td>500</td><td>0</td><td>500</td></tr>
<tr><td>P_2</td><td>I_3</td><td></td><td></td><td></td><td></td></tr>
<tr><td>1000</td><td>500</td><td>—</td><td>—</td><td>6·0 |500
0·5
0·0
6·0
12·0
———
24·5</td><td>6·0 |1000
0·5
2·0
6·0
10·0
———
24·5</td></tr>
<tr><td>2000</td><td>500</td><td>—</td><td>—</td><td>6·0 |500
0·5
0·0
0·0
12·0
———
18·5</td><td>6·0 |1000
0·5
2·0
0·0
10·0
———
18·5</td></tr>
<tr><td>1000</td><td>1000</td><td>—</td><td>2·5 |500
1·0
2·0
0·0
6·0
———
11·5</td><td>6·0 |1000
1·0
0·0
6·0
4·0
———
17·0</td><td>—</td></tr>
<tr><td>2000</td><td>1000</td><td>—</td><td>2·5 |500
1·0
2·0
6·0
6·0
———
17·5</td><td>6·0 |1000
1·0
0·0
0·0
4·0
———
11·0</td><td>—</td></tr>
<tr><td>1000</td><td>1500</td><td>2·5 |500
1·5
0·0
0·0
6·0
———
10·0</td><td>2·5 |1000
1·5
2·0
0·0
4·0
———
10·0</td><td>—</td><td>—</td></tr>
<tr><td>2000</td><td>1500</td><td>2·5 |500
1·5
0·0
6·0
6·0
———
16·0</td><td>2·5 |1000
1·5
2·0
6·0
4·0
———
16·0</td><td>—</td><td>—</td></tr>
</table>

tended to concentrate on methods of removing the seasonal influence, rather than determine low-cost production plans. For example, one method was to hire the kind of labor that would easily adapt to the seasonal requirements, such as farm labor for a product for which there is high demand in the winter. For three of the firms, he applied different planning methods and compared them to the normal operating schedule used by the firm in terms of annual variable costs. The results of the tests are shown in Table 4. The two methods used were linear decision rules and

Table 4

Annual Variable Costs of Alternative Scheduling Methods *

	Firms		
	A	*B*	*C*
1. Operating schedule	$1,616,023	$225,128	$255,988
2. Linear decision rule schedule	983,307	152,950	176,350
3. Simulation schedule	401,701	91,213	90,914
4. Level demand schedule	0	0	0
5. Total annual costs of firm	$54·5 millions	$9·0 millions	$16·5 millions

* The 'Operating Schedule' is the schedule developed by management and actually being used in the firm. From Vergin (1966).

simulation. Line 4 indicates a base assuming a completely level rate of demand. Then, the difference between the costs in lines 3 and 4 is attributable to seasonal factors, while the difference between lines 1 and 4 is due to both seasonal conditions and inefficient planning.

Figure 2 shows the general structure of a modular computer simulation program for the production planning problem, which is used to obtain the information in line 3 of Table 4. The planning process is essentially a trial-and-error procedure using guide rules developed by considering the cost figures and operating characteristics of the firm. Starting with a given plan, the effects of successively varying employment levels, overtime, inventory, subcontracting, etc., are evaluated successively until there is a convergence on a low-cost plan.

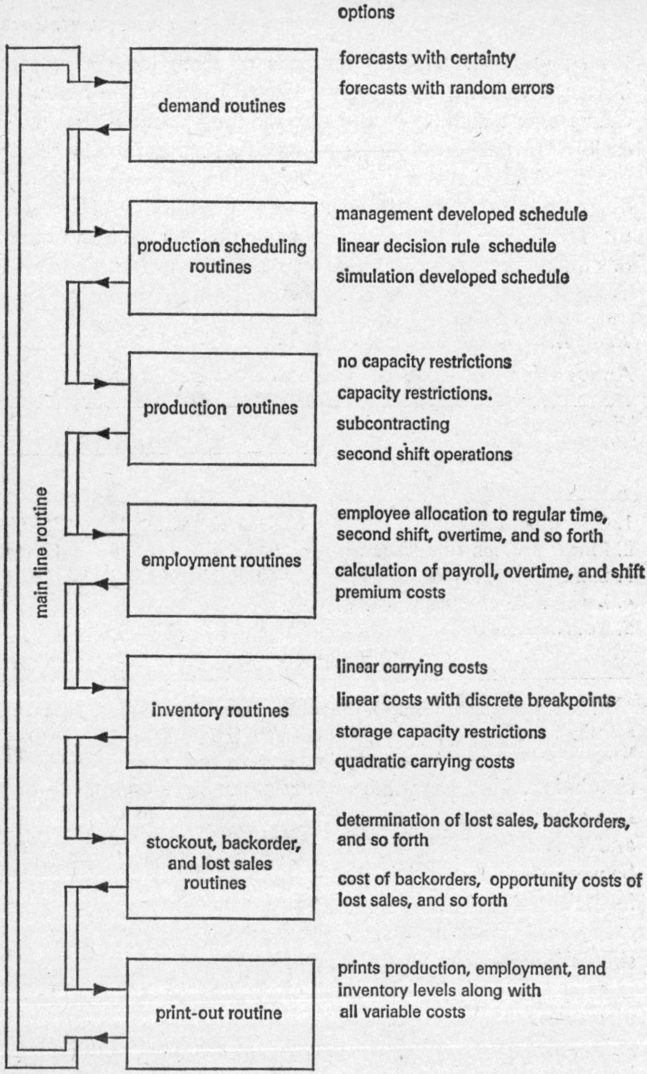

Figure 2 Flow diagram of simulation scheduling process. The main line routine sends controls of the program to each of the decision rule routines in turn. Each routine passes control back to the main line when it has performed its designated function. From Vergin (1966, p. 265)

Shop Loading

In the discussion of production planning, we generally avoided the problem of planning the production of individual items by considering production in the aggregate. If the final product mix consists of assemblies, then we are faced with the task of determining the production requirements for subassemblies and components and establishing purchasing schedules for raw materials with appropriate allowances for production time lags. Of course, these detailed schedules can be combined to form the basis of a general production plan. Our concern here will be with the task of synthesizing these data.

Alternative assembly plans

If the mix of final products consists of assemblies, then two basic approaches may be taken to meet assembly schedules. At the one extreme is the use of a work-in-process or subassembly and component inventory, which acts as a buffer between the final assembly department and the fabricating and subassembly operations. In this case, the requirements for subassemblies and components are met directly from this intermediate inventory. The orders for subassemblies, components, or materials, in turn, are triggered when it is necessary to replenish the inventory levels. Therefore, the demand schedule for final assemblies or completed parts is translated into production requirements through the control of the intermediate inventory system.

At the other extreme is a system of *product explosion* and *shop loading*. The demand for final assemblies is translated into an equivalent demand for subassemblies, components, and raw materials necessary to satisfy the final product demand. The requirements for subassemblies and component parts are then translated into hourly requirements for various production centers. Almost invariably this translation process must also consider the 'lead' times necessary to prepare the production orders, order materials, and process parts so that the subassemblies and components will arrive at the final assembly point when they are needed. In this system, an intermediate inventory may exist solely as a buffer for random fluctuations in demand. Product explosion and shop loading constitute the two aspects of this kind of program, the former being the translation of final assembly into component requirements, and the latter, the determination of the hours required in the shop to meet these demands.

Obviously, any plant may use a mixture of these two approaches. For example, suppose a company manufactures three lines of pleasure boats. Production planning might be confined to converting forecasted customer demand (distributors) into a schedule or plan for the assembly department. The woodworking shop, where the components or subassemblies are made, might operate on the basis of a production control plan under which shop orders are placed to replenish component inventories set up for the assembly lines. Or, the requirements for the final products might be translated directly into a plan for shop production with the necessary lead times taken into consideration.

Product explosion and subsequent shop loading can be effected either from a backlog of orders or from a forecast. If a system of backlogging orders is in effect, then as new orders arrive and are added to the backlog, they are exploded into the component requirements and translated into shop hour requirements, and the results are accumulated over time to provide a basic running schedule of future hourly requirements. The future shop load will depend on the lead time for the parts, as well as on the size of the order backlog.

In contrast to this is the possibility of forecasting the demand for final assemblies and then, again by explosion and shop loading, converting the forecast into a production schedule. The forecast period will have to be at least as long as the lead time necessary to complete the final product. Which scheme a company uses depends on the product, the company policies regarding customer service requirements, forecasting ability, etc.

Product explosion

Determining the requirements for subassemblies and components is not a trivial task, especially if parts and subassemblies are standardized for use in more than one final product. For large product mixes, the use of a computer is mandatory because of the magnitude of the arithmetic task. The mathematics of product explosion is essentially matrix algebra. We explore the subject by considering first a very simple problem.

Example 1. Suppose a company makes two final products, A and B, two subassemblies, a and b, and a basic component, x. The assembly relationships of the part, subassemblies, and final products are shown in Figure 3 in the form of a flow diagram (or

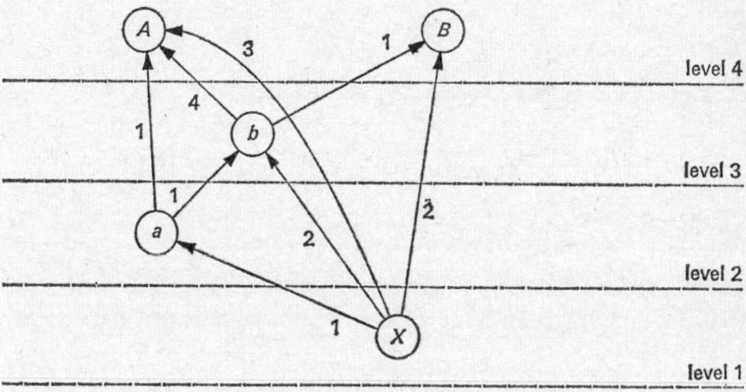

Figure 3 Flow diagram of assembly relationships between parts

gozinto chart). The arrows indicate the assembly relationship between two parts. The part at the tail of the arrow is used directly in the assembly of the part at the point of the arrow. The numbers next to the arrow are the quantities of the parts to be used in an assembly unit.

The parts are arranged in levels. Thus, a part located at one level may be used in any part in higher levels but never in a part at the same or lower level.

The N *matrix.* We now convert the information on the flow diagram into a matrix, $N = [n_{ij}]$, where n_{ij} is the quantity of the ith part (component or subassembly) used directly in the assembly of the jth part. The matrix contains exactly the same information found in the flow diagram but in a more useful form. Considering the matter of levels, we will code each part with a number, k. A part k always goes into parts with numbers greater than k but never into parts with numbers smaller than k.

	k	1	2	3	4	5
		j				
k	*i*	*x*	*a*	*b*	*A*	*B*
1	*x*	0	1	2	3	2
2	*a*		0	1	1	0
3	*b*			0	4	1
4	*A*				0	0
5	*B*					0

N matrix

The T *matrix.* The N matrix gives only the quantities of a part directly used in any other part. We are interested now in determining the total numbers of a part used directly and indirectly in any other part. For example, the number of *x*s used to subassemble one unit of *b* is three. *b* uses two *x*s directly and also one *x* contained in the *a* that goes into the *b*.

The T matrix gives us this information: $T = [t_{ij}]$, where t_{ij} is the quantity of the *i*th part required to assemble one unit of the *j*th part. T will be a square matrix with the same dimension as N.

The diagonal of the T matrix will be 1, or $t_{ii} = 1$. It can be shown that

$$T = I + N + N^2 + N^3 + \dots,$$

where I is the identity matrix. Since this is a geometric series of the form $a + an + an^2 + \dots + an^k$, an equivalent form is

$$T = \frac{I}{(I - N)},$$

since $$\lim_{k\to\infty} T = \frac{I}{(I - N)} = (I - N)^{-1}.$$

Therefore, T can be found by a matrix inversion operation. If the dimension of the N matrix is large, the inversion process can consume a good deal of computing time.

We may also calculate T by using the following algorithm:

$$t_{ij} = \sum_k (n_{ik} \,.\, t_{kj}) \quad \text{for} \quad i \leqslant k \leqslant j \quad \text{and} \quad t_{ii} = 1. \qquad 3$$

That is,

$$T = N \times T + I$$
$$[I \times T] - [N \times T] = I$$
$$[I - N][T] = I.$$

Therefore,

$$T = \frac{I}{I - N} = (I - N)^{-1}.$$

		1	2	3	4	5
		j				
	i	*x*	*a*	*b*	*A*	*B*
1	*x*	1	1	3	16	5
2	*a*	0	1	1	5	1
3	*b*	0	0	1	4	1
4	*A*	0	0	0	1	0
5	*B*	0	0	0	0	1

T matrix

It is fairly easy to systematically complete the T matrix given the N matrix. That is, suppose that we are given the N matrix shown at the left below and we wish to complete the first row of the T matrix indicated to the right of the N matrix. We proceed as follows:

$$0 \,.\, ? + 2 \,.\, 1 = 2$$
$$0 \,.\, ?' + 2 \,.\, 3 + 6 \,.\, 1 = 12$$
$$0 \,.\, ?'' + 2 \,.\, 7 + 6 \,.\, 2 + 8 \,.\, 1 = 34$$

$$\begin{bmatrix} 0 & 2 & 6 & 8 \\ & 0 & 3 & 1 \\ & & 0 & 2 \\ & & & 0 \end{bmatrix} \times \begin{bmatrix} 1 & ? & ?' & ?'' \\ & 1 & 3 & 7 \\ & & 1 & 2 \\ & & & 1 \end{bmatrix} + \begin{bmatrix} 1 & & & \\ & 1 & & \\ & & 1 & \\ & & & 1 \end{bmatrix} = \begin{bmatrix} 1 & 2 & 12 & 34 \\ & 1 & 3 & 7 \\ & & 1 & 2 \\ & & & 1 \end{bmatrix}$$

$$N \quad \times \quad T \quad + \quad I \quad = \quad T$$

Total period requirements. We may convert the requirements for a unit of end product into total requirements for some end product demand schedule. Suppose, for example, that we have a matrix, $S = [s_{ip}]$, where s_{ip} is the number of units of the ith part demanded in period *p*. We may convert this into a demand schedule for subassemblies and components as follows:

Let S =

i	1	2	3
A	6	4	5
B	5	3	2

(columns: *p*)

In order to obtain the total requirements of *x*, *a*, and *b* necessary to make the required number of *A* and *B* for each of the three periods, we carry out the following multiplication, using part of the T matrix:

i (rows) / *j*	*A*	*B*		*i* / *p*	1	2	3		*i* / *p*	1	2	3	Total
x	16	5	×	*A*	6	4	5	=	*x*	121	79	90	290
a	5	1		*B*	5	3	2		*a*	35	23	27	85
b	4	1							*b*	29	19	22	70

Shifting the time base. Usually, components and subassemblies must be produced in advance of the time they are actually required. For instance, suppose that part *b* must be subassembled one period prior to its use in *A* and *B*. This means that all *x*s and *a*s going into *b* must be prepared in the same period or, more likely, in a preceding period. In other words, each level of parts constitutes an increment of lead time between the point where components are processed (or materials ordered) and the final assembly of *A* and *B*.

For purposes of demonstration, we will assume that the *x*s and *a*s can be produced in the period for which they are demanded. Using the N matrix, we can determine the distribution of this demand over the subassemblies and components required in the third period.

i / *j*	*x*	*a*	*b*	*A*	*B*		*i* / *p*	3		*i* / *j*	*x*	*a*	*b*	*A*	*B*
x	0	1	2	3	2	×	*x*	90	=	*x*	0	27	44	15	4
a	0	0	1	1	0		*a*	27		*a*	0	0	22	5	0
b	0	0	0	4	1		*b*	22		*b*	0	0	0	20	2
A	0	0	0	0	0		*A*	5		*A*	0	0	0	0	0
B	0	0	0	0	0		*B*	2		*B*	0	0	0	0	0

This shows that, in terms of the third period demand, 19 *x*s are required directly for *A* and *B*, 44 are required directly for the *b*s, which will be made in period 2, and 27 are required for the *a*s that go directly into *A* and will be used in period 2 for the *b*s. An entire production schedule for four periods is shown below and can be verified by extending the analysis to periods 1 and 2.

Period	*Part* *A*	*B*	*a*	*b*	*x*
3	5	2	5 for direct use in *A*	0	19 direct in *A* and *B*; 5 direct in *a*s;
2	4	3	4 direct in *A* 22 direct in *b*s	22 for period 3 *A* and *B*	18 direct *A* and *B*; 4 direct *a*s for *A*; 44 direct in *b*s; 22 direct in *a*s for *b*s;
1	6	5	6 direct in *A* 19 direct in *b*s	19 for period 2 demand	28 direct *A* and *B*; 6 direct *a*s for *A*; 38 direct for *b*s; 19 direct *a*s for *b*s;
1-1	0	0	29 direct in *b*s	29 for period 1 demand	58 direct in *b*s; 29 direct in *a*s for *b*s.
	Total:		**85**	**70**	**290**

Loading the facilities

Suppose we have four facilities in the above example. Designating a facility by the letter *m*, we have an A matrix, $A = [a_{im}]$, where a_{im} is the time to process one unit of part *i* in facility *m*.

	i \ *m*	1	2	3	4
	x	1	2	1	0
	a	0	1	2	3
A =	*b*	0	0	3	5
	A	0	0	6	5
	B	0	0	7	6

Hours required to produce one unit of part *i* in facility *m*

To determine the total hourly requirements for each of the four facilities for each of the four periods, we summarize the production schedule in the form of a total parts requirements matrix:

p \ i	x	a	b	A	B
3	24	5	0	5	2
2	88	26	22	4	3
1	91	25	19	6	5
1-1	87	29	29	0	0

Total part requirement for four periods

Then, the total hourly requirement is calculated as follows, giving us an H matrix, H = $[h_{pm}]$, where h_{pm} is the number of hours required for the mth facility in the pth period:

p \ i	x	a	b	A	B
3	24	5	0	5	2
2	88	26	22	4	3
1	91	25	19	6	5
1-1	87	29	29	0	0

×

i \ m	1	2	3	4
x	1	2	1	0
a	0	1	2	3
b	0	0	3	5
A	0	0	6	5
B	0	0	7	6

=

p \ m	1	2	3	4
3	24	53	78	52
2	88	202	251	226
1	91	207	269	230
1-1	87	203	232	232

Generalizing the problem with constant lead times

In the preceding example, we derived a schedule of part requirements, considering that only part *b* was subjected to a lead time. Suppose we generalize the problem by saying that all final products will be processed in the month they are demanded and all components and subassemblies will be processed in one time period prior to the time they are needed.

Let

$u =$ the number of final products
$r =$ the number of components and subassemblies
$p_t =$ the number of periods for which a forecast of final product shipments is made, $t = 1, 2, \ldots, \mathrm{T}$, where T is the most advanced period
$\mathrm{N}^* =$ a submatrix from the N matrix
$S =$ a shipment schedule
$D =$ a parts requirement schedule.

Then we have the following computation:

$$r \overset{u}{\boxed{\quad \mathrm{N}^* \quad}} \times u \overset{\substack{p_t \\ \mathrm{T}, \ldots, 1}}{\boxed{\quad S \quad}} = r \overset{\substack{p_t \\ \mathrm{T}-1, \ldots, \mathrm{T}-\mathrm{T}}}{\boxed{\quad D_1 \quad}}$$

D_1 is the requirements for the r parts to meet the shipment schedule S, but to be made in period $\mathrm{T} - 1, \mathrm{T} - 2, \ldots, \mathrm{T} - \mathrm{T}$.

$$r \overset{r}{\boxed{\quad \mathrm{N}^* \quad}} \times r \overset{\mathrm{T}-1, \ldots, \mathrm{T}-\mathrm{T}}{\boxed{\quad D_1 \quad}} = r \overset{\mathrm{T}-2, \ldots, \mathrm{T}-(\mathrm{T}+1)}{\boxed{\quad D_2 \quad}}$$

D_2 is the requirements for the r parts to meet the requirements schedule D_2, but to be made one period in advance of the requirements D_1. In this manner, we continue until we generate a requirement schedule, $D = 0$.

Now we combine the requirement schedules thus derived into a total schedule of production for each of the parts for each of the periods, T, $\mathrm{T} - 1$, $\mathrm{T} - 2, \ldots$. Given this schedule, we proceed to load the facilities in the manner previously described. We should note that a check on the part requirements can be made by comparing the total number of parts with the product of $\mathrm{T} \times S$.

Estimating shop load requirements by regression

In a job shop, where each part is made to customer specification, it may be difficult to estimate the hourly requirements of individual parts for various facilities because of the variety of parts produced. However, the experience gained in the past may be used to predict or estimate requirements for a new part by using multiple linear regression analysis.

For example, this method has been used by the Westinghouse Electric Corporation to estimate labor time requirements for building up the laminations that form the core of a shell-type power transformer (Canning, 1959, pp. 447–49). An engineering drawing of the part is provided several months before production time. For this kind of part, the following variables, each a dimension of the part, were considered to contribute to the core building time.

X_1 = number of stacks
X_2 = yoke area (in^2)
X_3 = short area (in^2)
X_4 = leg area (in^2)
X_5 = total weight of iron (k lb)
X_6 = $\frac{1}{6}$ opening volume (k in^3)
X_7 = $\frac{1}{2}$ core volume (k in^3)
X_8 = stack height (in).

By examining a sample of 400 parts produced in the past, each with varying dimensions, and using multiple linear regression analysis, the following estimating formula was developed:

$$Y = T_0 + T_1 X_1 + T_3 X_3 + T_5 X_5 + T_7 X_7 + T_8 X_8,$$

where Y is the estimated time and T_i is the coefficient derived from the sample. Note that a preliminary study of a random sample of 50 resulted in the elimination of variables X_2, X_4, and X_6 as being redundant.

A test of this estimating equation on a sample of 100 current shop orders showed that 30 per cent of the estimated times were within ± 1 per cent of the actual times, 75 per cent were within ± 5 per cent, and the maximum error, which occurred in 2 per cent of the cases, was ± 20 per cent.

This method has been successfully used in other shops of the company and has provided a rudimentary method of scheduling.

Summary

We have considered how a set of resources is to be allocated in the aggregate over a future planning period to meet a variable demand. This demand is either a backlog, a forecast, or both. We have considered the resources to be employment levels, facility hours, over- and undertime, and inventories (accrued capacity) or backlogs (deferred capacity).

The degree of aggregation depends on the extent of product and facility mix. The typical planning model requires the translation of heterogeneous product demand into equivalent demand for production capacity. Both product explosion and shop loading are procedures for making this translation. When this is done, planning is concerned with the economics of satisfying the capacity requirements, given constraints on some of the capacity factors for each future period. We should be cognizant of the fact that the design and implementation of production plans can be a formidable task. The effectiveness of optimum-seeking methods depends on the reliability of forecasts, the economics of data collection, the accuracy of cost estimates, and the ability of the model to entertain the relevant planning variables. Plans never represent a firm commitment for the future but must be updated periodically during the planning year. Simulation appears to offer the ultimate in flexibility in terms of ability to experiment with alternative planning policies and to consider a wide variety of planning variables.

The next two chapters will consider the more tactical problems of operating the production system on a short-term basis in terms of individual products and facilities. The control of production at the operational level is executed within or without the framework of aggregate planning. The latter case exists in those situations where forecasting is difficult, product and facility mixes are large, and relatively minor increments and decrements in capacity area are easy to make on a short-term basis. When planning has been carried out, a problem of tactical production control is to reconcile operating decisions with constraints set forth by aggregate planning. That is, the actual aggregate result of decisions on the size of production batches, which product inventories are to be replenished, and how production is to be scheduled in the shop must be reconciled with the desired or planned aggregate inventory levels, anticipated shop loads, and employment and overtime levels.

References

CANNING, F. L. (1959), 'Estimating road requirements in a job shop', *J. Indust. Eng.*, vol. 10, no. 6.

HOLT, C. C., MODIGLIANI, F., MUTH, J. F., and SIMON, H. A. (1960), *Planning Production, Inventories, and Work Force*, Prentice-Hall.

HOLT, C. C., MODIGLIANI, F., and SIMON, H. A. (1955), 'A linear decision rule for production and employment scheduling', *Management Science*, vol. 2, no. 1.

MAGEE, J. F., and BOODMAN, D. M. (1967), *Production Planning and Inventory Control*, McGraw-Hill.

VERGIN, R. C. (1966), 'Production scheduling under seasonal demand', *J. Indust. Eng.*, vol. 17, no. 5.

14 E. H. Bowman and R. B. Fetter

Uses of Transportation Procedure

Excerpt from E. H. Bowman and R. B. Fetter, *Analysis for Production and Operations Management*, 3rd edn, Richard D. Irwin, 1967, pp. 134–6.

Uses of Transportation Procedure

A number of types of problems can be placed within the framework of the transportation method of linear programming. Some of these are as follows:

1. Conventional transportation problems in which a number of sources, such as plants, supply a number of destinations, such as warehouses.
2. Machine and job assignment problems in which a number of jobs can be done on a number of machines or by a number of people.
3. Scheduling over several time periods.

These types of applications are mainly suggestive, rather than inclusive. It is the nature of the problem that is important; if it can be considered a source–destination or input–output problem conforming to the attributes discussed here, it might be possible to place it within the framework of the transportation method of linear programming. These attributes particularly linearity, homogeneity, and certainty, while not strictly met, may be sufficiently well approximated to permit use of the method.

The time period problem might be explained further. The sales experience of many manufacturers follows a seasonal pattern. Either production can be constant with fluctuating inventories or inventories can remain constant with fluctuating production. Some combination of the two is also possible. This problem may sometimes be conceived as a problem of balancing overtime costs (fluctuating production) with storage costs (fluctuating inventories). In this case, each production possibility (regular time and overtime) in each time period may be considered a

Table 1

Product	5/10	6/10	8/10	10/10	12/10	14/10	16/10	18/10	20/10	22/10	24/10	26/10	28/10	30/10	*Capacity in hrs*
(1) A	5454 F	5095 F	6025 F	6025 F	4600 F	4125 F	4125 F	3958 F	3980 F	4175 F	4430 F	4430 F	–	–	600 h
(2) B	10,422 F	9637 F	9333 F	8620 F	8540 F	8500 F	8440 F	8420 F	8380 F	8310 F	8310 F	8270 F	8270 F	8270 F	600 h
(3) C	–	–	–	12,430 F	11,010 F	10,400 F	10,150 F	10,150 F	–	–	–	–	–	–	400 h
Planned production	5000 t	2200 t	5100 t	5000 t	4200 t	3000 t	2000 t	1300 t	750 t	500 t	500 t	250 t	100 t	100 t	
A	20·7	23·8	40	40	30	40	40	45·3	44·5	38·7	36	36	–	–	
B	7·2	8·6	9·3	11·5	11·8	12	12·2	12·3	12·5	12·8	12·8	13	13	13	
C	–	–	–	9	12	14	15	15	–	–	–	–	–	–	
(*Output, tons per hr*)															
	788	2200	5100	5000	389	3000	2000	1300	750	500	500	250	–	–	
	4212	0	0	0	0	0	0	0	0	0	0	0	100	100	
	–	–	–	0	3811	0	0	0	–	–	–	–	–	–	

Assignments according to the optimum (rounded to the ton). Minimum total cost = 199,822,700 F. Unused capacity: A = B = 0; C = 82·5h.

source of input. Each sales requirement (i.e. product group) in each time period may be considered a separate destination or output. The nature of the costs in the 'cell routes' in the table is a combined production cost (regular or overtime) plus a storage cost for the appropriate number of time periods. A general table for this problem is shown in Table 1 (Bowman, 1956).

Virtually half such a transportation matrix is crossed out, as it is not possible to deliver in a period previous to that made or shipped. This facilitates a more rapid solution.[1]

Many transportation problems, such as plants and warehouses, can be extended to include the time period problem where this is meaningful. If requirements of the warehouses fluctuate during the year, as in the case, for example, of a seasonal pattern, the problem becomes one of determining whether to use the more favorable routes (revealed by the regular transportation method) during slack periods and incur storage costs or use the less favorable – more expensive – routes during peak periods and not incur the storage costs. This problem may be placed within the above time period framework by considering each plant in each time period a separate source and each warehouse in each time period a separate destination. Each cost figure in the matrix will be a combination of manufacturing, transportation, and storage costs. It is also possible to extend this same method to the multiproduct case.

1. In effect, this formulation of the problem places an infinite cost on supplying sales in one period from subsequent periods. In many cases, 'back ordering' may be possible. Some finite cost must then be placed on this time lag. One way of looking at this problem is to consider it as a combination of storing inventory and/or storing customer orders, both processes having their own costs.

References

Bowman, E. H. (1956), 'Production scheduling by the transportation method of linear programming', *Operations Research*, vol. 4.

15 A. Kaufmann

Study of a Monthly Production Plan *and* Choice of Energy Sources in Electrical Power Generation

Excerpts from A. Kaufmann, *Methods and Models of Operations Research*, Prentice-Hall, 1963, pp. 36–9, 44–9.

Study of a Monthly Production Plan

We have three rolling mills, A, B, and C. We plan to manufacture, as cheaply as possible, fourteen classes of plate, of the following thicknesses: 5/10, 6/10, 8/10, 10/10; 12/10, 14/10, 16/10, 18/10, 20/10, 22/10, 24/10, 26/10, 28/10, and 30/10. We know the cost per ton of each class of plate in each of the mills. We also know the output in tons per hour for each plate thickness, the capacity of each mill, and the production forecasts. The capacities and the forecasts are given for one month.

Table 1 presents:

1. The cost per ton c_{ij} of plate j in mill i (in francs).
2. The output d_{ij} of plate j in mill i (in tons per hour).
3. The production p_j planned for plate j (in tons).
4. The upper limit b_j of time available in mill i (in hours).

The problem is to find the minimum total monthly production cost.

Table 2 presents the linear program for the model of the problem. The mathematical constraints involve 14 equations, 3 inequations, and 31 variables. By introducing 3 slack variables, λA, λB, and λC, to represent the unused capacity, we obtain a system of 17 equations and 34 variables. All basic solutions must contain at least $34 - 17 = 14$ null variables.

Simply by examination of Table 1, some interesting intuitive conclusions are immediately apparent. First of all, mill A has the lowest costs and yields the highest hourly output; thus, we should assign the largest possible number of tons in our plan to this mill. The 600-hour limitation on the capacity of mill A means that considerable work will have to be assigned to mills B and C. This is the point at which difficulties arise and we begin to

Table 1

Production Scheduling by the Transportation Method

Sales periods (destination)

		Sales$_1$	*Sales*$_2$	*Sales*$_3$	... *Sales*$_n$	*Inventory*$_n$	*Slack*	*Total capacities*
	Inventory$_0$	0	C_I	$2C_I$	... $(n-1)C_I$	nC_I	0	I_O
Production	Regular$_1$	C_R	$C_R + C_I$	$C_R + 2C_I$	... $C_R + (n-1)C_I$	$C_R + nC_1$	0	R_1
Periods	Overtime$_1$	C_O	$C_O + C_I$	$C_O + 2C_I$	... $C_O + (n-1)C_I$	$C_O + nC_I$	0	0_1
(*Source*)	Regular$_2$	–	C_R	$C_R + C_I$	... $C_R + (n-2)C_I$	$C_R + (n-1)C_I$	0	R_2
	Overtime$_2$	–	C_O	$C_O + C_I$	... $C_O + (n-2)C_I$	$C_O + (n-1)C_I$	0	0_2
	Regular$_3$	–	–	C_R	... $C_R + (n-3)C_I$	$C_R + (n-2)C_I$	0	R_3
	Overtime$_3$	–	–	C_O	... $C_O + (n-3)C_I$	$C_O + (n-2)C_I$	0	0_3
	...	...	...	...	... ...	...	...	...
	Regular$_n$	–	–	–	... C_R	$C_R + C_I$	0	R_n
	Overtime$_n$	–	–	–	... C_O	$C_O + C_I$	0	0_n
	Total Require-ments	S_1	S_2	S_3	... S_n	I_n	*	

* Slack total = $I_0 + R + 0 - S - I_n$.

Notation: I_i = Inventory at the end of the ith time period. R_i = Maximum number of units which can be produced during ith time period on regular time. O_i = Maximum number of units which can be produced during ith time time period on over-time. S_i = Number of units of finished product to be sold (delivered) during ith time period. C_R = Cost of production per unit on regular time. C_O = Cost of production per unit on overtime. C_I = Cost of storage per unit per time period.

appreciate that intuition, like rule-of-thumb reasoning, is simply not adequate to determine how the extra work should be assigned. Calculation becomes indispensable. For this example, calculations were performed according to Dantzig's method, or rather according to a variation of that method. A gamma-type magnetic-drum computer was used. The computation took 10 minutes and required 21 iterations. Done by hand, it would have taken a good man several months.

The optimum program is shown at the bottom of Table 2. Intuition would certainly not have dictated an assignment of 4212 tons of 5/10 plate to mill B, where the cost is highest (10,422 francs), or an assignment of 3811 tons of 12/10 plate to mill C.

If we compare a common program solution, worked out by intuition or on the basis of experience, with the optimum program, we find that the latter provides a saving of 6 per cent. Even if the difference had been slighter, it would have been very important to know the optimum.

Choice of Energy Sources in Electrical Power Generation [1]

For an enterprise like Electricité de France, working out a production plan consists in minimizing an economic function that includes the total cost of investments plus the operating expenses adjusted by the market interest rate, while at the same time satisfying consumer power demand. Satisfying this demand involves meeting specified conditions for the following three dimensions:

1. Guaranteed power-average hourly consumption during the heavy business hours in wintertime, in megawatts (Mw), i.e. thousands of kilowatts.
2. Peak power-average hourly consumption during the four daily peak hours of the same wintertime business days, in Mw.
3. The annual power consumption, in gigawatt-hours (Gwh), i.e. millions of kilowatt-hours.

To satisfy consumer demand, one can utilize various types of power plants:

steam (thermal), hydroelectric (river-current, reservoir, dam, etc.), tidal-power, or nuclear power. We assume, to simplify

1. See Gibrat (1955); Massé (1955); Ventura (1956).

Table 2

Mathematical Constraints

$$\frac{x_{11}}{20{\cdot}7} + \frac{x_{12}}{23{\cdot}8} + \frac{x_{13}}{40} + \frac{x_{14}}{40} + \frac{x_{15}}{30} + \frac{x_{16}}{40} + \frac{x_{17}}{40} + \frac{x_{18}}{45{\cdot}3} + \frac{x_{19}}{44{\cdot}5} + \frac{x_{1\cdot10}}{38{\cdot}7} + \frac{x_{1\cdot11}}{36} + \frac{x_{1\cdot12}}{36} \leqslant 600$$

$$\frac{x_{21}}{7{\cdot}2} + \frac{x_{22}}{8{\cdot}6} + \frac{x_{23}}{9{\cdot}3} + \frac{x_{24}}{11{\cdot}5} + \frac{x_{25}}{11{\cdot}8} + \frac{x_{26}}{12} + \frac{x_{27}}{12{\cdot}2} + \frac{x_{28}}{12{\cdot}3} + \frac{x_{29}}{12{\cdot}5} + \frac{x_{2\cdot10}}{12{\cdot}8} + \frac{x_{2\cdot11}}{12{\cdot}8} + \frac{x_{2\cdot12}}{13} + \frac{x_{2\cdot13}}{13} + \frac{x_{2\cdot14}}{13} \leqslant 600$$

$$\frac{x_{34}}{9} + \frac{x_{35}}{12} + \frac{x_{36}}{14} + \frac{x_{37}}{15} + \frac{x_{38}}{15} \leqslant 400$$

$$x_{11} + x_{21} = 5000$$
$$x_{12} + x_{22} = 2200$$
$$x_{13} + x_{23} = 5100$$
$$x_{14} + x_{24} + x_{34} = 5000$$
$$x_{15} + x_{25} + x_{35} = 4200$$
$$x_{16} + x_{26} + x_{36} = 3000$$
$$x_{17} + x_{27} + x_{37} = 2000$$
$$x_{18} + x_{28} + x_{38} = 1300$$
$$x_{19} + x_{29} = 750$$
$$x_{1\cdot10} + x_{2\cdot10} = 500$$
$$x_{1\cdot11} + x_{2\cdot11} = 500$$
$$x_{1\cdot12} + x_{2\cdot12} = 250$$
$$x_{2\cdot13} = 100$$
$$x_{2\cdot14} = 100.$$

31 variables
17 equations or inequations

Economic function

$$z = 5454\,x_{11} + 5095\,x_{12} + 6025\,x_{13} + 6025\,x_{14} + 4600\,x_{15} + 4125\,x_{16} + 4125\,x_{17} + 3958\,x_{18} + 3980\,x_{19} + 4175\,x_{1\cdot10} + 4430\,x_{1\cdot11}$$
$$+ 4430\,x_{1\cdot12} + 10{,}422\,x_{21} + 9637\,x_{22} + 9333\,x_{23} + 8620\,x_{24} + 8540\,x_{25} + 8500\,x_{26} + 8440\,x_{27} + 8420\,x_{28} + 8380\,x_{29} + 8310\,x_{2\cdot10}$$
$$+ 8310\,x_{2\cdot11} + 8270\,x_{2\cdot12} + 8270\,x_{2\cdot13} + 8270\,x_{2\cdot14} + 12{,}430\,x_{34} + 11{,}010\,x_{35} + 10{,}400\,x_{36} + 10{,}150\,x_{37} + 10{,}150\,x_{38}.$$

The optimum

$z = 199{,}822{,}700$ francs.

matters, that all plants of a given type constitute a homogeneous series – in other words, show the same characteristics. This will give us the steam-plant series, the hydroelectric series, etc. For the typical plant in each series (series i for example), we shall define:

a_i = guaranteed power,
b_i = peak power,
c_i = total annual power,
d_i = investment cost,
f_i = annual operating expenses,
x_i = number of plants[2] of the series-*i* type.

We then construct the following basic hypothesis: the quantities a_i, b_i, c_i, d_i, and f_i are constants. This hypothesis is practically acceptable. Further, let us define:

A = total guaranteed power (for the entire system comprising all the plants, no matter what their type),
B = peak power output (for the entire system),
C = total annual power output.

A production plan must meet the following three conditions or constraints:

$$a_1 x_1 + a_2 x_2 + \ldots + a_n x_n \geqslant A \qquad \mathbf{1}$$
$$b_1 x_1 + b_2 x_2 + \ldots + b_n x_n \geqslant B \qquad \mathbf{2}$$
$$c_1 x_1 + c_2 x_2 + \ldots + c_n x_n \geqslant C. \qquad \mathbf{3}$$

To these technical constraints we shall add a fourth financial constraint: let D be the ceiling on investment expenditures. This will give us:

$$d_1 x_1 + d_2 x_2 + \ldots + d_n x_n \leqslant D. \qquad \mathbf{4}$$

Lastly, by introducing the slack variables x_a, x_b, x_c, and x_d, the inequations from **1–4** can be written:

$$\left.\begin{aligned} a_1 x_1 + a_2 x_2 + \ldots + a_n x_n - x_a &= A \\ b_1 x_1 + b_2 x_2 + \ldots + b_n x_n - x_b &= B \\ c_1 x_1 + c_2 x_2 + \ldots + c_n x_n - x_c &= C \\ d_1 x_1 + d_2 x_2 + \ldots + d_n x_n + x_d &= D \end{aligned}\right\} \qquad \mathbf{5}$$

2. We can relate the plants by a unit of output in megawatts, and an actual plant will embody a given number of such units.

The economic function will include:
The initial investments,

$$d_1 x_1 + d_2 x_2 + \ldots + d_n x_n. \qquad 6$$

The annual cost of installation maintenance, which will bring into play a coefficient of capitalization k. If e_i denotes the annual maintenance cost for a plant in series i, we obtain for the total maintenance cost

$$k(e_1 x_1 + e_2 x_2 + \ldots + e_n x_n). \qquad 7$$

The annual cost of coal for the steam plants. We shall adopt the index 1 to represent these plants; let us call c'_1 the annual power supplied by a steam plant, and p_0 the cost of coal per thermal kilowatt-hour produced. We note that actually c'_1 will differ from c_1 because the steam plant does not operate for the total number of hours per year that its guaranteed power output would allow. Hence the annual cost of coal will be $kp_0 c'_1 x_1$.

In fact, we choose c_1 in such a way that[3]

$$c_1 x_1 = c_1 x_1 - x_c, \qquad 8$$

where x_c is the margin of power saved each year in steam.

Hence $$c'_1 x_1 = C - c_2 x_2 - c_3 x_3 - \ldots - c_n x_n. \qquad 9$$

Therefore, the economic function will be

$$z = \sum_{i=1}^{n} d_i x_i + k \sum_{i=1}^{n} e_i x_i + k p_0 C - \sum_{i=2}^{n} c_i x_i. \qquad 10$$

The annual operating costs f_i will then be presented by

$$f_1 = e_1, \quad f_2 = e_2 - p_0 c_2, \quad \ldots, \quad f_n = c_n - p_0 c_n. \qquad 11$$

Let us now define

$$g_1 = d_1 + kf_1, \quad g_2 = d_2 + kf_2, \quad \ldots, \quad g_n = d_n = kf_n. \qquad 12$$

The economic function will then be written

$$z = g_1 x_1 + g_2 x_2 + \ldots + g_n x_n + kp_0 C. \qquad 13$$

Taking $$z' = z - kp_0 C, \qquad 14$$

3. The c'_1/c_1 ratio is called the rate of steam utilization for the year.

we will then attempt to find the minimum value of

$$z' = g_1 x_1 + g_2 x_2 + \ldots + g_n x_n \qquad \mathbf{15}$$

while the n variables must satisfy the four constraints **5**:

$$\begin{aligned} a_1 x_1 + a_2 x_2 + \ldots + a_n x_n - x_a &= A \\ b_1 x_1 + b_2 x_2 + \ldots + b_n x_n - x_b &= B \\ c_1 x_1 + c_2 x_2 + \ldots + c_n x_n - x_c &= C \\ d_1 x_1 + d_2 x_2 + \ldots + d_n x_n + x_d &= D. \end{aligned} \qquad \mathbf{16}$$

The linear program (**15** and **16**) contains $n + 4$ variables; a possible solution, in order to be optimum, must contain at least n null variables. Thus, since there are only 4 constraints, an optimum solution can contain at most 4 means of production. This fact, which is not intuitive, is very important indeed; it shows that by adopting a fifth constraint (as, for example, a social constraint), we can find an optimum that can contain an extra non-null variable, which means an extra means of production.

To make the problem practical, let us use some actual figures. In 1955, the French Parliament set these goals:

A (guaranteed power)	1692 Mw
B (peak power)	2307 Mw
C (annual power output)	7200 Gwh

and as a constraint:

D (investment ceiling): an arbitrary sum (a value of *D* constitutes a policy).

Table 3

	a_i	b_i	c_i	d_i	g_i
				in millions of francs	
1. Steam plant	1	1·15	7	97	136
2. River-current plant	1	1·10	12·6	420	56
3. Reservoir plant	1	1·20	1·3	130	101
4. Dam plant	1	3	7·35	310	104
5. Tidal-power plant	1	2·13	5·47	213	79

Table 3 gives the specifications for the 5 types of plant. If, in addition, we have the following data:

1. Interest rate: 8 per cent, or $k = 12{\cdot}5$.
2. Cost of fuel for 1 thermal kwh: 3 francs. We obtain the sum of 270 billion francs as the value of $kp_0\, C$.

The linear program to be optimized is then

$$\left.\begin{aligned} z &= 136x_1 + 56x_2 + 101x_3 + 104x_4 + 79x_5 \\ x_1 + x_2 + x_3 + x_4 + x_5 - x_a &= 1692 \\ 1{\cdot}15x_1 + 1{\cdot}10x_2 + 1{\cdot}2x_3 + 3x_4 + 2{\cdot}13x_5 - x_b &= 2307 \\ 7x_1 + 12{\cdot}6x_2 + 1{\cdot}3x_3 + 7{\cdot}35x_4 + 5{\cdot}47x_5 - x_c &= 7200 \\ 97x_1 + 420x_2 + 130x_3 + 310x_4 + 213x_5 + x_d &= D. \end{aligned}\right\} \quad 17$$

Let us select as a possible basic solution the one that corresponds to

$$x_2 = x_3 = x_4 = x_5 = x_b = 0 \qquad 18$$

or, in other words, a solution where only steam power is used, and in which the total peak power load is met. This yields

$$\begin{aligned} x_1 &= 2006 \qquad & x_a &= 314 \\ x_c &= 6842 \qquad & x_d &= D - 194{,}582. \end{aligned} \qquad 19$$

It appears that such a solution requires an investment that must exceed 195 billion francs (rounding off 194,582).

With a solution in which

$$x_1 = x_3 = x_4 = x_5 = x_b = 0, \qquad 20$$

we would obtain

$$x_2 = 2097 \quad x_a = 405 \quad x_c = 19{,}222 \quad x_d = D - 880{,}740, \quad 21$$

in which case D must be greater than 881 billion francs (rounded off).

It is possible to make a comparative study of all the basic solutions. There are, at the most, 126 basic solutions in this program (we do well to emphasize 'at the most', since among these 126 solutions, some will contain negative values of the variables and must be discarded). The basic solutions can include one, two, three, or four types of active plants.

The reader will find, in the works of Massé and Gibrat as well as Ventura, a very complete and particularly interesting discussion of this problem.

Using an electronic computer, or even by paper-and-pencil calculations, we can study the variation in the optimum of z for different values of D, and examine, for each of these values, which kinds of plants should be used.

The considerations treated in this section make it possible to solve the basic problem of the theory of plans. This problem can be stated as follows: What modifications can we make, either in the objectives of a plan or in the costs of the various types of plants available for the formulation of a plan, in order that the optimum solution may keep its structure?

References

GIBRAT, R. (1955), 'Le problème général des plans de production d'énergie électrique et les usines marémotrices', *Électricité de France Document.*

MASSÉ, G. (1955), *Nouvelle Revue d'Économie Contemporaine*, February.

VENTURA, E. (1956), 'Un exemple de recherche opérationnelle: la determination d'un plan optimum de production d'énergie électrique par la théorie des programmes linéaires', SOFRO Document.

16 M. K. Starr and D. W. Miller

Multiple-Item Inventories

Excerpt from M. K. Starr and D. W. Miller, *Inventory Control: Theory and Practice*, Prentice-Hall, 1962, pp. 93–104.

More than One Item: The Optimal Policy Curve

Let us consider the basic optimal lot-size model extended to the case where there are a number of items for which inventories must be maintained. Very few organizations have only one item which they must maintain in inventory and, as organizations get larger, the trend is very decidedly in the direction of enormous numbers of items which must be maintained in stock. Some organizations number their stock-keeping units in the hundreds of thousands. We want to see whether the existence of a multitude of items changes the problems, adds new problems, or increases the range of possible solutions to the problems.

On first view there appears to be no reason to expect any changes simply because of the existence of large numbers of items in inventory. The equation for the optimal amount to order contains no term representing number of items in inventory. Further, it is evident that the same equation can be applied to each item, using the data for that item, and that the optimal amount to order for the item will be determined by the equation. The sum of the optimal policies for each item will represent the best overall solution to the total inventory problem. Therefore, there is no reason to expect any change in our analysis resulting from the hypothesized large number of items.

All of these remarks are true but, first, there are some new problems which can arise and, second, there are some new possibilities which arise for the resolution of the inventory problems. It is this which we shall be mainly considering in this section.

The new possibilities arise in connexion with the problem of measuring the costs required by the optimal lot-size equation.

There can be some serious difficulties in the way of successfully estimating these costs. The worst of these difficulties stems from the fact that we really must measure opportunity costs, and cost accounting data is not typically suitable for this purpose. We will simply develop some means for avoiding these problems which exist in the case of inventories of many items.

We need an illustrative example. Suppose that a company maintains inventories of five items. The yearly demand, z_i, and the cost per unit, c_i, for these items are:

Item	z_i	c_i
1	600	$3
2	900	10
3	2400	5
4	12,000	5
5	18,000	1

Typically, if one were called upon to make a study of this company's inventory problem one would find that the company was using some ordering system such as ordering each item once a month. Assuming that this is the case, we can calculate the number of orders per year and the average inventory which results:

Item	*Orders per year*	*Average inventory*
1	12	$75
2	12	375
3	12	500
4	12	2500
5	12	750
	60	$4200

where the average inventory is easily calculated from $(z_i c_i)/24$.

If, now, we are told that the ordering cost for this company is $10 per order and the carrying cost is 12 per cent per year, we can calculate the total cost of the company's current ordering policy,

$$T\,C_{\text{Current}} = 60(10) + 0{\cdot}12(4200) = \$1104.$$

Knowing that $C_r = \$10$ and that $C_c = 0{\cdot}12$, we can easily calculate the optimal ordering policy for each of the items. We will use the expression for order size in terms of dollars rather than in terms of units

$$x_{i\$} = \sqrt{\left(\frac{2c_i\, z_i\, C_r}{C_c}\right)}.$$

Using this expression we can quickly calculate:

Item	*Optimal order size in dollars*	*Orders per year*
1	\$548	3·28
2	1225	7·35
3	1414	8·49
4	3162	18·98
5	1732	10·39
	\$8081	48·49

where orders per year are calculated directly from $(z_i\, c_i)/x_{i\$}$. Since the average inventory for each item is simply one-half the order size in dollars we can immediately determine the total cost of the ordering policy we have calculated:

$$\mathrm{T\,C_{opt}} = 10(48{\cdot}49) + 0{\cdot}12(4040) = \$970.$$

We will not consider here the question of rounding off the order periods but will assume that the company can equally well use any desired order period for an item. It will be seen that the utilization of our equation for the optimal policy has resulted in a decrease in total costs of \$134, or about 12 per cent.

Before we continue the intended direct line of discussion let us take a sidetrack and consider a kind of problem which might arise but which would not require any analysis in the case of one item. Suppose that the company has a shortage of working capital and cannot afford to maintain the indicated optimal average inventory investment of \$4040. Precisely, let us suppose that the most the company can afford in inventory investment is an average \$3000. How should the ordering policy be changed to minimize the total costs under this restriction? Let us formulate the problem mathematically. Using our summation notation, the total cost equation when x_i units of the ith item are ordered is

$$\mathrm{T\,C} = 10 \sum_i \frac{z_i}{x_i} + 0{\cdot}12 \sum_i \frac{x_i\, c_i}{2}.$$

We have been given the restriction that

$$\sum_i \frac{x_i\, c_i}{2} = 3000.$$

Our problem is to minimize T C subject to the given restriction. The mathematical technique for accomplishing this is to use Lagrangian multipliers. This rather formidable sounding technique is actually very easy to use. We write the restriction as an equation equal to zero

$$\sum_i \frac{x_i\, c_i}{2} - 3000 = 0.$$

Now we form the Lagrangian, denoted L, by adding λ, the multiplier, times this equation to the expression to be minimized, TC. Thus, we have

$$L = 10 \sum_i \frac{z_i}{x_i} + 0{\cdot}12 \sum_i \frac{x_i\, c_i}{2} + \lambda \left(\sum_i \frac{x_i\, c_i}{2} - 3000 \right).$$

To minimize the original expression subject to the restriction we must minimize L over the x_i and λ. We do this by the usual procedure of taking derivatives and setting them equal to zero and we get

$$\frac{\partial L}{\partial x_i} = -\frac{10 z_i}{x_i^2} + 0{\cdot}06 c_i + \frac{\lambda c_i}{2} = 0,$$

$$\frac{\partial L}{\partial \lambda} = \sum_i \frac{x_i\, c_i}{2} - 3000 = 0.$$

These simultaneous equations can be readily solved by determining x_i from the first equation:

$$x_i = \sqrt{\left[\frac{20 z_i}{(0{\cdot}12 + \lambda) c_i} \right]},$$

substituting this in the second equation:

$$\sqrt{\left(\frac{20}{0{\cdot}12 + \lambda} \right)} \sum_i \sqrt{(z_i\, c_i)} = 6000$$

and solving for λ:

$$\lambda = 20 \left[\frac{\sum \sqrt{(z_i\, c_i)}}{6000} \right]^2 - 0{\cdot}12.$$

In order to obtain a numerical value for λ we must determine the square of the sum of the square roots of $z_i c_i$ from the original data. We have

Item	$\sqrt{(z_i c_i)}$
1	42·43
2	94·87
3	109·54
4	244·95
5	134·16
	625·95

From this we immediately calculate

$$\lambda = 0{\cdot}09767.$$

Now we can determine the x_i from the equation above which includes λ. Even more conveniently we can calculate

$$x_{i\$} = \sqrt{\left(\frac{20}{0{\cdot}12+\lambda}\right)}\sqrt{(z_i\,c_i)}$$
$$= 9{\cdot}585\sqrt{(z_i\,c_i)}$$

since we have already determined $\sqrt{(z_i\,c_i)}$. Thus, we obtain

Item	*Order size in dollars*	*Orders per year*
1	$406·70	4·43
2	909·30	9·90
3	1049·90	11·43
4	2348·00	25·56
5	1286·00	14·00
	$5999·90	65·32

The total cost for this ordering policy is

$$\text{TC} = 10(65{\cdot}32) + 0{\cdot}12(3000) = \$1013.$$

This is the minimum cost ordering policy of all possible policies which would give an average inventory investment of $3000. It will be seen that the total cost has gone up as compared to the optimal policy – as it obviously must. However, for an increase in total costs of only $43 the company has accomplished its purpose of cutting its average inventory investment by $1040.

Our sidetrack has fulfilled the subsidiary purpose of introducing the extremely useful device of Lagrangian multipliers as a way to minimize expressions subject to restrictions. We will need this device again when we return to our main theme. Suppose, now, that there are grievous difficulties in the way of measuring the two costs necessary to determine the optimal ordering policy for the company. Our analysis of the optimal ordering policy subject to the requirement that the average dollar inventory should not exceed \$3000 suggests an approach which might be used. The existing company policy produces an average inventory of \$4200 and uses 60 orders per year. Now, with no knowledge of C_r or C_c whatsoever, we can certainly reason that if we can reduce inventory and maintain the same number of orders per year then we have improved the company's position. This must be true, regardless of the costs involved. We propose, then, to see if we can decrease the average inventory investment while keeping the number of orders per year fixed at 60. If we can do this we will have improved the situation even though we will not be able to determine the actual amount of the saving.

We will proceed exactly as before. We want to minimize the inventory subject to the restriction that total orders per year will be 60. The expression for total average inventory is

$$\text{Total average inventory} = \text{TI} = \sum_i \frac{x_i c_i}{2}.$$

We want $$\text{Total orders} = \text{TO} = \sum_i \frac{z_i}{x_i} = 60.$$

We form the Lagrangian

$$L = \sum_i \frac{x_i c_i}{2} + \lambda\left(\sum_i \frac{z_i}{x_i} - 60\right),$$

and set the derivatives equal to zero:

$$\frac{\partial L}{\partial x_i} = \frac{c_i}{2} - \frac{\lambda z_i}{x_i^2} = 0,$$

$$\frac{\partial L}{\partial \lambda} = \sum_i \frac{z_i}{x_i} - 60 = 0.$$

Solving the first equation for x_i gives

$$x_i = \sqrt{\left(2\lambda \frac{z_i}{c_i}\right)}$$

and substituting this in the second equation and solving for λ gives

$$\lambda = \frac{\left[\sum_i \sqrt{(z_i c_i)}\right]^2}{7200}.$$

We already have determined $\sum_i \sqrt{(z_i c_i)}$ for these data so we calculate directly

$$\lambda = 54{\cdot}42.$$

From this we now obtain

$$x_{i\$} = x_i c_i = \sqrt{(2\lambda z_i c_i)} = 10{\cdot}43\sqrt{(z_i c_i)}.$$

This gives:

Item	*Order size in dollars*	*Orders per year*
1	\$442·70	4·07
2	989·80	9·09
3	1142·80	10·50
4	2554·90	23·48
5	1399·30	12·86
	\$6529·50	60·00

The inventory is one-half of the order size in dollars or \$3265. The number of orders has remained at 60, our restriction, and the inventory has been reduced by \$935, or 22·3 per cent of its amount under the current policy. This is the minimum possible inventory investment with 60 orders per year. The decrease in inventory investment has been achieved by redistributing the orders among the items in an optimal way. We cannot, of course, determine what this reduction in inventory is worth to the company because we do not know the value of C_c.

The success of this analysis immediately suggests the corresponding argument in reverse. Why not improve the company's position, without knowledge of the costs, by keeping the average inventory investment at the same level and decreasing the total number of orders per year? This can also be done. The expressions for TI and TO are the same as before, but their roles are reversed. We want to minimize

$$\mathrm{TO} = \sum_i \frac{z_i}{x_i}$$

subject to $$\mathrm{TI} = \sum_i \frac{x_i\, c_i}{2} = \$4200.$$

We form the Lagrangian,

$$L = \sum_i \frac{z_i}{x_i} + \lambda\left(\sum_i \frac{x_i\, c_i}{2} - 4200\right)$$

and set the derivatives equal to zero

$$\frac{\partial L}{\partial x_i} = -\frac{z_i}{x^2} + \frac{\lambda c_i}{2} = 0,$$

$$\frac{\partial L}{\partial \lambda} = \sum_i \frac{x_i\, c_i}{2} - 4200 = 0.$$

We find $$x_i = \sqrt{\left(\frac{2z_i}{\lambda c_i}\right)}$$

$$\lambda = \frac{\left[\sum_i \sqrt{(z_i\, c_i)}\right]^2}{35{,}280{,}000}.$$

For our data $$\lambda = 0{\cdot}01111$$

and $$x_{i\$} = 13{\cdot}42\sqrt{(z_i\, c_i)}.$$

Therefore, we can calculate:

Item	*Order size in dollars*	*Orders per year*
1	\$569·40	3·16
2	1273·10	7·07
3	1469·90	8·16
4	3286·40	18·26
5	1800·00	10·00
	\$8398·80	46·65

Average inventory remains at \$4200, within rounding errors, while total orders per year have been reduced to 46·65, a saving of 13·35 orders per year or 22·3 per cent of the orders under the current policy. This policy is the one which has the minimum total orders per year of all policies which would give an average inventory investment of \$4200. It will be noted that the percentage saving is the same in both cases, 22·3 per cent. This will always be the case. The question of which of these two possible improved policies is preferable cannot be answered without knowledge of the relevant costs. However, either of them will clearly save the company money – even though we cannot calculate how much.

Now we have calculated three different optimal policies under different restrictions and we calculated the original optimal policy for assumed values of C_r and C_c. Careful inspection of these four different optimal policies indicates an underlying uniformity among them. This is shown in the following table:

Policy	*Total Inventory*	*Orders per year*	*Product* (TI)(TO)
Optimal; $C_r = \$10$, $C_c = 0{\cdot}12$	\$4040	48·49	195900
Optimal; Inventory = \$3000	3000	65·32	195960
Optimal; Orders = 60	3265	60	195900
Optimal; Inventory = \$4200	4200	46·65	195930

All the products of TI times TO are equal to a constant value, within rounding errors. Further, with a few tries it is possible to discover where this constant value, 195,900, originates. More precisely, the constant value is 195,910 and it is equal to

$$\tfrac{1}{2}(625{\cdot}95)^2 = \tfrac{1}{2}\Big[\sum_i \sqrt{(z_i\, c_i)}\Big]^2.$$

We must certainly undertake to explain the reason for this structural similarity among the various optimal policies.

Reviewing the equations for $x_{i\$}$ which we used in determining the actual optimal ordering policies in each of the four cases we see that in each of them we had

$$x_{i\$} = K\sqrt{(z_i\, c_i)}.$$

K, of course, differed in the four cases but it was always determined either from the costs, C_r and C_c, or from λ. Presumably this is the reason for the occurrence of $(\sum \sqrt{z_i\, c_i})^2$ in the constant product of TI and TO. However, instead of pursuing this line of reasoning we will take a different route which will lead us to the same goal. Let us write the expression for TI in terms of the optimal $x_{i\$}$, as determined in the optimal lot-size model,

$$\mathrm{TI} = \sum_i \frac{X_{i\$}}{2} = \sum_i \frac{1}{2}\sqrt{\left(\frac{2c_i\, z_i\, C_r}{C_c}\right)} = \sqrt{\left(\frac{C_r}{C_c}\right)} \sum_i \sqrt{\left(\frac{c_i\, z_i}{2}\right)}.$$

Similarly, the expression for TO is

$$\mathrm{TO} = \sum_i \frac{z_i\, c_i}{x_{i\$}} = \sqrt{\left(\frac{C_c}{C_r}\right)} \sum_i \sqrt{\left(\frac{c_i\, z_i}{2}\right)}.$$

These two expressions are valid for any values of C_r and C_c whatsoever – plugging in any given values and the data on $z_i\, c_i$ will give the optimal total inventories and total orders per year. Now, the interesting thing about these two expressions is that when they are multiplied together the two costs, C_r and C_c, cancel out and we are left with

$$(\mathrm{TI})(\mathrm{TO}) = \tfrac{1}{2}\left[\sum_i \sqrt{(z_i\, c_i)}\right]^2,$$

exactly the constant product we already discovered empirically from our four optimal policies. Thus, we see that the product of TI and TO, for the optimal policy based on any possible values of the two costs, will always be a constant for the given data. Further, we see that if we divide our two expressions everything cancels out except the two costs:

$$\frac{\mathrm{TI}}{\mathrm{TO}} = \frac{C_r}{C_c}.$$

Therefore, given the TI and TO for any optimal policy we can immediately determine the ratio of the two costs upon which the optimal policy was constructed.

This last fact suggests an alternative interpretation of the optimal policies we determined under various restrictions. When we restricted the inventory to \$3000 we found the optimal policy had 65·32 orders per year and, of course, \$3000 in inventory. The ratio, (TI)/(TO), is therefore, for this policy equal to 45·9. If the order cost is known to be \$10 we can conclude that this policy is imputing $C_c = \$10/45{\cdot}9 = 0{\cdot}218$. In other words, the policy we determined would only be optimal if $C_c = 0{\cdot}218$. Therefore, the imputed value of C_c must be 0·218. Now, this policy resulted because the company had a shortage of working capital and wanted to restrict their inventory investment to \$3000. We have discovered that this restriction automatically imputes a value of 0·218 to the carrying cost. We can reason similarly for the other two optimal policies under restrictions. Restricting orders to 60 and assuming, for example, that the carrying cost is 0·12 automatically imputes a value of \$6·53 to C_r. Restricting inventory to \$4200 and assuming that the order cost is \$10 automatically imputes a value of 0·111 to the carrying cost.

The relation between these imputations and the various λs we calculated is an interesting one. For our first restriction we assumed that we knew that $C_r = \$10$ and $C_c = 0{\cdot}12$ but that the company was forced to restrict the inventory to \$3000. We found $\lambda = 0{\cdot}09767$ and we note that $0{\cdot}12 + 0{\cdot}0977 = 0{\cdot}218$, the imputed carrying cost in this case. In other words, λ is not an arbitrary mathematical device. Its value is equal to the increment added to the carrying cost because of the inventory restriction. It plays, in short, the role of a cost in this formulation. The second restriction we imposed was that orders should equal 60. We found $\lambda = 54{\cdot}42$ and it will be noted that

$$54{\cdot}42\,(0{\cdot}12) = \$6{\cdot}53,$$

the imputed value of C_r for $C_c = 0{\cdot}12$. Therefore, λ in this case is the ratio C_r/C_c. The third restriction was that inventory should equal \$4200. We found

$$\lambda = 0{\cdot}01111 \text{ and } 10(0{\cdot}01111) = 0{\cdot}111$$

is exactly the imputed carrying cost for this case. Here in other words, λ is the ratio C_c/C_r. In every case, λ is intimately related to the imputed costs resulting from restrictions. In most economic problems in which it is necessary to use Lagrangian multiplier methods it develops similarly that λ has a very basic economic significance.

Returning to the original policy of the company we have another way of indicating its irrationality. In every one of our optimal policies we can take any one of the items individually and it will have the given ratio between its orders per year and its average inventory. As a matter of fact, this is the essence of a rational inventory policy in a nutshell – making every item's policy consistent in this respect. Now, if we look at the current policy of the company we will find that there is no such uniformity among the individual items. This has nothing to do with the fact that the company policy isn't optimal for its own specific costs. It is that the policy could not be optimal for any conceivable single set of costs. For the company's current policy we can use our argument on the individual item policies and discover the imputed carrying costs assuming that the order cost is actually $10:

Item	(TI)/(TO)	*Imputed* C_c
1	6·25	1·60
2	31·25	0·32
3	41·67	0·24
4	208·33	0·05
5	62·50	0·16

The wide range in these imputed values is an indication of the irrationality of the company's current policy.

Finally, our argument provides the means for a device which can be most useful to the executive responsible for establishing inventory policies under circumstances where it is difficult or impossible to obtain satisfactory estimates of the relevant costs. This device is called the optimal policy curve. The basis of this curve is precisely the constant product, (TI)(TO), which we have found holds true, for a given set of data, for the optimal policy for any possible combination of values of C_r and C_c. We can make a graph of this curve, using total inventory investment as one axis and total orders per year for the other axis. The equation, $(TI)(TO) = K$, is that of a hyperbola. Figure 1 shows this hyperbola, the optimal policy curve, for the data of our example. Every point on this curve represents the optimal policy for some ratio of C_r to C_c, specifically, the ratio for any point on the curve is given by (TI)/(TO). Further, every possible optimal policy is represented by a point on this curve.

By means of this curve we can recapitulate our argument. Point A is the current company policy. Its irrationality is indicated by the fact that it does not fall on the curve. When we argued that we could improve the company position by holding

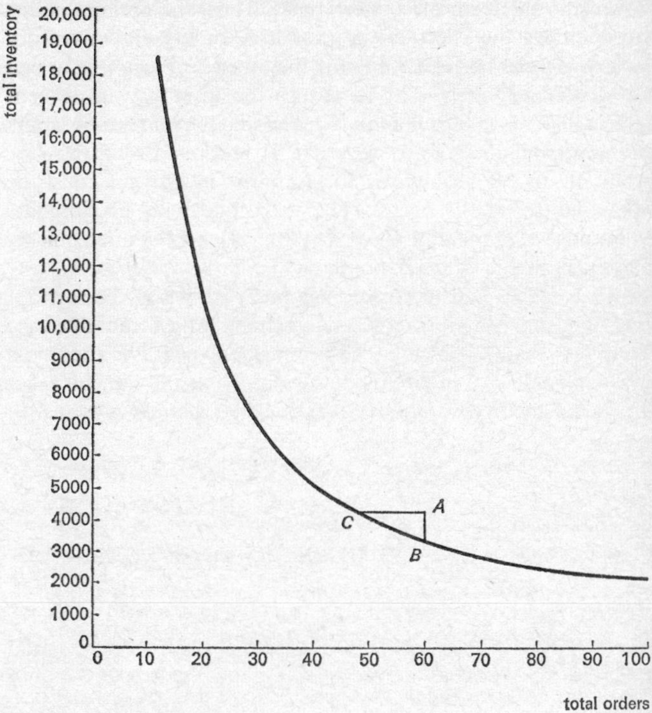

Figure 1 Optimal policy curve

orders at 60 and reducing the inventory investment, we were simply saying that by running a parallel to the TI axis from point A until it intersects the optimal policy curve, at B, we must be improving the situation. When we held the inventory fixed at $4200, we ran a parallel to the TO axis from A until it intersected the optimal policy curve at C. This, obviously, must also improve the company position.

However, the optimal policy curve can be considerably more useful than simply as a geometrical analogy to an already completed mathematical argument. It provides means for the executive to utilize his knowledge and experience under circumstances where this knowledge and experience cannot be boiled down into the form of cost estimates. Thus, the optimal policy curve shows the executive exactly how orders and inventory investment can be traded one for the other. For example, it can be seen immediately that to reduce the inventory investment from $3000 to $2000 will require increasing the number of orders per year from about 65 to about 99. To decrease orders per year from 40 to 30 requires that inventory investment must be increased by about $1600. The executive, with his intimate knowledge of the circumstances of the company, can often quickly converge on the optimal point on the curve for the company without having had to convert his knowledge into the form of carrying and ordering costs – something which can often be done only badly if at all. The optimal policy curve, then, is a most valuable tool in the frequently occurring and difficult cases where satisfactory estimates of the relevant costs are not available.

17 R. G. Brown

Inventory Safety Factors

Excerpts from R. G. Brown, *Statistical Forecasting for Inventory Control*, McGraw-Hill, 1959, pp. 105–27, 200–202.

Every man must judge for himself between conflicting vague possibilities. *Darwin*

These doles that you are now distributing neither suffice to ensure your safety.... *Demosthenes*

One of the basic questions in inventory control can be answered with the help of the classical theory of economical order quantities. The answer to the other question depends on an estimate of the maximum reasonable demand during a lead time. One way of estimating the maximum reasonable demand is to estimate the expected demand and then to add an allowance for protection against the uncertainty inherent in any forecast. The allowance is the product of the safety factor and the standard deviation of the errors in forecasting over a lead time.

The safety factor is the control that reflects what you consider *reasonable* in protecting service to your customers. In one approach to the problem, you will have to decide when the marginal cost of providing more inventory is greater than the marginal value of giving better service. In another approach, the safety factor can be selected to balance the cost of carrying the inventory against a known cost of having backorders.

We shall take up first the methods of selecting a safety factor for one particular item, under each of these criteria. The latter sections will deal with means for estimating total investment and average service for all items in your product line.

1. Safety Factors for Specified Service Percentage

When you sell a product, you know its *cost* (not always with great precision but with accuracy that depends on the eagerness and ability of your accounting department). The *price* is a policy decision; generally it is above the cost, to give you a net profit, but there are occasions when it is strategic to offer to sell an item below cost.

When you buy a product, you know the *price*. Its *value* to you is now a policy decision. Generally speaking, if the value is greater than the price (and if you have the money), you will buy the item. If the value is below the price, you won't buy it.

The 'product' I want to take up now is a concept – the service rendered to your customers by your inventory. You can, of course, give service in many other ways in addition to having stock on the shelf when the order comes in. You can expedite deliveries of stock on order; you can offer the customer an acceptable substitute; you can try to convince him that it would take even longer to get the item from a competitor, and so on. These other opportunities and the price you have to pay for using them must all be taken into account when you decide the value of giving service by routinely having the stock on hand when the demand comes in.

The value is subjective; it changes with time, depending on the service rendered by competitors and on your need or desire to increase or decrease the general level of demand. The marginal value of giving better service probably also changes. If you can fill only 75 per cent of the demand from stock, it may be worth a great deal to increase the routine service to 80 per cent. However, if you're giving 98 per cent service already, there may not be much value for increasing it to 99 per cent – you'd like to, of course, but you won't pay much for it.

Therefore, my approach will be to show how to compute the price you have to pay, in terms of inventory investment, for increasing the service from any level. When you know how much you'll have to pay to increase your service one, five, or ten percentage points from the present level, you can decide when the increase in price is greater than the increase in value to you and set your safety factors to give that particular level of service routinely.

(a) *Periodic replenishment reviews*

Let's take up first the system in which you review the stock at fixed intervals and order enough to bring the available stock back up to the estimated maximum reasonable demand during a lead time. Remember that in this system, the lead time includes not only the procurement lead time but also one interval equal to the time between reviews.

(The estimate of maximum reasonable demand is also reviewed periodically. In many systems, the estimate is reviewed at the same time the inventory levels are reviewed. It is not necessary to review the estimates that often. You could, for example, determine the estimate of maximum reasonable demand once a quarter but review stock levels against that estimate every month. Many companies review stocks at the end of each day's business against a forecast made monthly.)

Let us assume that we have an estimate of the average rate of demand during the lead time and also an estimate of the standard deviation of the forecast errors over the lead time. We also assume, as a starting point, that the forecast errors are normally distributed.

How many standard deviations should be added to the expected demand to give an estimate of the maximum reasonable demand during the lead time? In other words, what safety factor should you choose? In what follows we shall need some measure of the service rendered routinely from stock on hand. I shall use the amount of demand filled from stock, expressed as an average fraction of the total demand. (An alternative might be the fraction of the time there are backorders.)

If the demand is less than the maximum reasonable level decided on, all of it will be filled. If the demand is greater than the maximum reasonable level, only an amount equal to the estimated maximum reasonable level will be filled; the excess demand will be backordered.

The first step in finding the appropriate safety factor for any given service percentage is to compute the value of a 'service

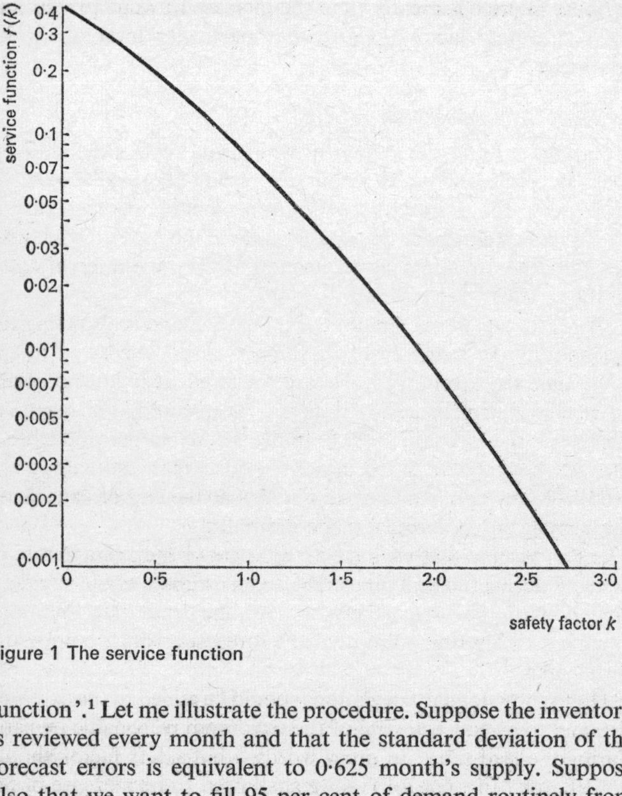

Figure 1 The service function

function'.[1] Let me illustrate the procedure. Suppose the inventory is reviewed every month and that the standard deviation of the forecast errors is equivalent to 0·625 month's supply. Suppose also that we want to fill 95 per cent of demand routinely from stock. The service function is

$$f(k) = \frac{1}{0{\cdot}625}(1 - 0{\cdot}95) = 0{\cdot}08.$$

1. This is a mathematical function, not a job to be done in the shipping department. It is defined by

$$f(k) = \int_k^\infty (t-k)\,p(t)\,dt = p(k) - k\int_k^\infty p(t)\,dt,$$

where $p(t)$ may be assumed to be the normal probability density function. It is the average quantity backordered (measured in standard deviations). (Brown, 1957.)

The general formula is

$$f(k) = \frac{Q}{M}(1 - P),$$

where Q is the number of months between successive reviews of the inventory, M is the standard deviation of the forecast errors (through whatever the lead time is), expressed in months' supply, and P is the desired service percentage, expressed as a decimal.

Finally, look up a value for the safety factor from Figure 1 or the corresponding Table 1. In our example, the value of the service function is 0·08, which corresponds to a safety factor of $k = 1{\cdot}0$.

Table 1

The Service Function for a Normal Distribution of Forecast Errors

Safety factor k	*Service function* $f(k)$
0·0	0·3989
0·2	0·3069
0·4	0·2304
0·6	0·1687
0·8	0·1202
1·0	0·0833
1·2	0·0561
1·4	0·0367
1·6	0·0232
1·8	0·0143
2·0	0·0085
2·2	0·0049
2·4	0·0027
2·6	0·0015
2·8	0·0008
3·0	0·0004

Figure 2 is designed to make it easier to find the safety factor directly from the ratio Q/M and the service percentage and illustrates more clearly how the safety factor changes when one or more of the other variables change. For example, a specific

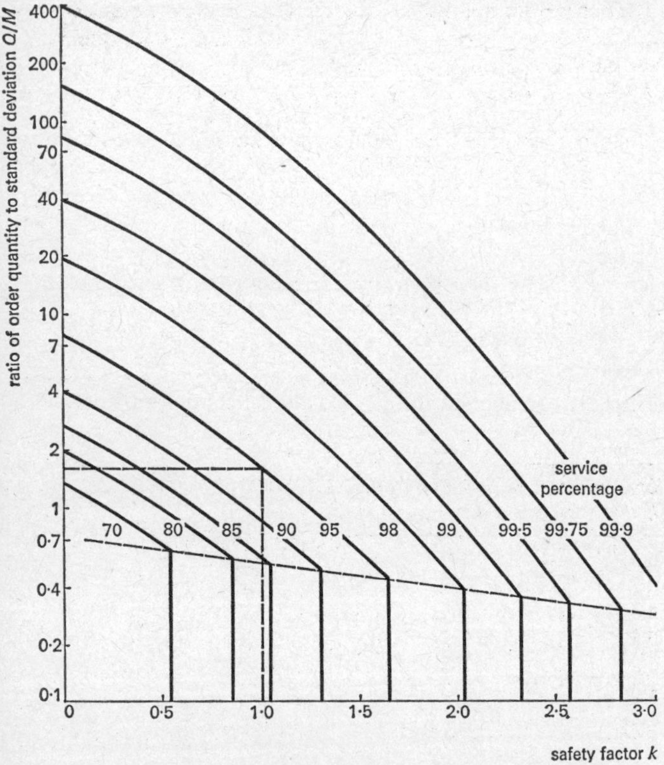

Figure 2 Direct-reading chart for safety factor

item has a definite ratio Q/M, such as the value $1/0{\cdot}625 = 1{\cdot}6$; therefore, as you move horizontally along the line corresponding to 1·6 in Figure 2, you can see how the safety factor must change as the desired service level changes. If you want 98 per cent service, the safety factor will increase to about 1·5.

Note that when the order quantity is small in comparison with the standard deviation of forecast errors, the cycle stock does not provide any significant protection, and the safety factor can be determined directly from the normal distribution (Table 2). If the standard order quantity is less than about one-half the standard deviation of forecast errors (depending on the level of service desired), it is not important in providing stock for protection of service. As the order quantity increases, the need for safety stock becomes less and less. It will be typical to find that slow-moving items may be near the top of the chart and that faster movers, which are ordered in smaller time supplies, are near the bottom.

Table 2

The Cumulative Normal Distribution

F(*t*) is the probability that an error will be smaller than *t* standard deviations.

t	*F*(*t*)	*t*	*F*(*t*)	*t*	*F*(*t*)
0·00	0·5000	1·00	0·8413	2·00	0·9773
0·05	0·5199	1·05	0·8531	2·05	0·9798
0·10	0·5398	1·10	0·8643	2·10	0·9821
0·15	0·5596	1·15	0·8749	2·15	0·9842
0·20	0·5793	1·20	0·8849	2·20	0·9861
0·25	0·5987	1·25	0·8944	2·25	0·9878
0·30	0·6179	1·30	0·9032	2·30	0·9893
0·35	0·6368	1·35	0·9115	2·35	0·9906
0·40	0·6554	1·40	0·9192	2·40	0·9918
0·45	0·6736	1·45	0·9265	2·45	0·9929
0·50	0·6915	1·50	0·9332	2·50	0·9938
0·55	0·7088	1·55	0·9394	2·55	0·9946
0·60	0·7258	1·60	0·9452	2·60	0·9953
0·65	0·7422	1·65	0·9595	2·65	0·9960
0·70	0·7580	1·70	0·9554	2·70	0·9965
0·75	0·7734	1·75	0·9599	2·75	0·9970
0·80	0·7881	1·80	0·9641	2·80	0·9974
0·85	0·8023	1·85	0·9678	2·85	0·9978
0·90	0·8159	1·90	0·9713	2·90	0·9981
0·95	0·8289	1·95	0·9744	2·95	0·9984

There are two commonly followed policies. The first is to set a target level of service for all items and then to choose safety factors for each one from Figure 2. The other is to choose one safety factor for all items, with the intent of providing better-than-average service routinely on the slow movers. These policies are discussed more fully in section 3, p. 277.

There is a special case of this type of system which deserves some comment. Suppose you receive a regular replenishment shipment, either once a week or whenever your total order fills a truckload. In these cases you would review the stocks of all items at the same interval and order (perhaps with some limits imposed by normal handling quantities) what you had used of each one.

If the standard deviation of usage can actually be expressed as a multiple of the average usage, then the ratio Q/M will be the same for all items, and one value of the safety factor will provide enough stock for you to give the same average service on all items.

(b) *Review after each transaction*

In the other basic type of inventory-control system, the available stock (stock on hand, plus stock on order, less any unfilled demand) is reviewed after each transaction. If the demand reduces the available stock below an order point, an order is released to replenish stock. The economical quantity to order (as well as the economical interval between reviews in the first type of system) is covered exhaustively elsewhere.

The order point should be equal to an estimate of the maximum reasonable demand during a lead time. And again, we shall express that estimate as

Maximum reasonable demand = expected demand +
+ (safety factor) × (standard deviation).

The only difference between this and the previous system is the definitions given to the various terms. We still compute a value for the service function, but with two changes. Instead of letting Q represent the number of months between reviews, we let it represent the standard order quantity (in months' supply) that is released whenever stock is to be replenished. And the standard deviation of the forecast errors is now to be computed for only the standard replenishment lead time; the lead time does not include an interval between reviews.

Let's see what happens to the safety factor as each of the three component factors – standard order quantity, forecast accuracy, and desired service – changes.

Standard order quantity. The standard order quantity generates some inventory. Just after the shipment is received, the inventory is high. Just before the next one is received, it is low. On the average, the low point is equal to the safety stock, although, of course, sometimes it will be as low as zero, and sometimes it will be higher than the safety stock. (If the low-point inventory didn't fluctuate above and below the safety stock, there wouldn't be any reason for having safety stock!)

On the average, the inventory will be equal to the safety stock plus half of a standard order quantity.

There is no problem in giving good service just after a shipment has been received: stocks are high, and any demand can be filled promptly. The only time there is any danger of not being able to meet the demand is just before the next shipment is received. If the order quantity is large, there will be few shipments per year and therefore few opportunities to backorder.

Looking at Figure 2, we see that a larger standard order quantity would have us enter at a higher point on the vertical scale, which would lead to a smaller safety factor for the same service. Therefore we can conclude that an increase in the size of the standard order quantity will decrease the size of the safety stock, and the total inventory will not, in general, increase by half of the increase in the order quantity.

Forecast accuracy. If you are following my advice, you are designing a forecasting system that will be fairly accurate, so that the standard deviation of the forecast errors should be relatively low. A low standard deviation also means a low safety factor. Thus, with an accurate system of forecasting, the standard deviation is low, and you need fewer standard deviations – a double saving in safety-stock investment.

The forecast accuracy is expressed as the standard deviation of error in the lead time. Therefore, a decrease in the procurement lead time will also result in a lower safety stock. For example, suppose that your procurement lead time is now six months and that the standard deviation of forecast error is 1·5 months' supply. If the standard order quantity is a three-month supply and you want 95 per cent service, the service function will be

$$f(k) = \frac{3}{1{\cdot}5}(1 - 0{\cdot}95) = 0{\cdot}10,$$

which is equivalent to a safety factor of 0·9. The safety stock will then be (0·9) × (1·5) = 1·35 months' supply.

If you could cut the procurement lead time from six to four months, the standard deviation would be cut approximately in proportion to 1·0 month's supply, which would be equivalent to a safety factor of 0·65, and the safety stock would drop to (0·65) × (1·0) = 0·65 months' supply. Cutting the lead time by a third reduces the *total* investment from 1·5 + 1·35 = 2·85 months' supply to 1·5 + 0·65 = 2·15 months' supply, or by about a quarter. (The safety-stock requirement would be cut in half.)

Service desired. If you want a larger fraction of demand to be filled routinely from stock, the inventory level of course has to be raised. How much? The service function and the safety factor give us the answer.

Let's see first what service you can expect to give with no safety stock at all. Just after a shipment has been received, you can give perfect service for a while. As the stock level falls, at some time you will release another order to replenish stocks, but in half of those times, you will have run out before the new shipment is received. The value of the service function for a zero safety factor is 0·3989. Therefore, the fraction of the demand filled from stock without delay will be

$$P = 1 - 0{\cdot}3989\,\frac{M}{Q}.$$

If we continue our example where the standard replenishment quantity is $Q = 3$ months' supply and the standard deviation of forecast errors is $M = 1{\cdot}5$ months' supply, we find that

$$P = 1 - 0{\cdot}3989\,\frac{1{\cdot}5}{3} = 0{\cdot}80.$$

That is, with no safety stock at all, we would fill 80 per cent of the demand from the shelf!

Now, what would it cost to improve the service? Work Sheet 1 is designed to help compute the answer. In the first column, list several alternatives for the desired service. (If you use the service levels plotted in Figure 2 you can skip columns 2 and 3 and read the safety factors directly.) The second column shows the factor $(1 - P)$. Since in our example, the ratio of the standard order quantity to the standard deviation of forecast errors is $Q/M = 2$, the service function, in the third column, is just twice the second

Worksheet 1

Computing Investment versus Service

Standard replenishment quantity $Q = 3$ (*months*). *Cycle stock* $Q/2 = 1{\cdot}5$ (*months*)

Standard deviation of forecast errors $M = 1{\cdot}5$ (*months*) $Q/M = 2{\cdot}0$

1	*2*	*3*	*4*	*5*	*6*
Desired service	*Comple-ment*	*Service function*	*Safety factor*	*Safety stock*	*Total stock*
P	$1-P$	$\frac{Q}{M}(1-P)$	k	kM	$\frac{Q}{2}+kM$
0·80	0·20	0·40	0·00	0·00	1·50
0·85	0·15	0·30	0·20	0·30	1·80
0·90	0·10	0·20	0·50	0·75	2·25
0·925	0·075	0·15	0·65	0·98	2·48
0·95	0·050	0·10	0·90	1·35	2·85
0·96	0·040	0·08	1·00	1·50	3·00
0·97	0·030	0·06	1·15	1·73	3·23
0·98	0·020	0·04	1·35	2·03	3·53
0·99	0·010	0·02	1·68	2·52	4·02
0·995	0·005	0·01	1·95	2·93	4·43

column. The safety factor can then be read from Figure 1 and recorded in the fourth column. The safety stock is the standard deviation (1·5 months' supply) times the safety factor. Finally, the total inventory is the safety stock plus half the standard order quantity. The results can be plotted as in Figure 3, with a smooth curve to help determine the price you have to pay for intermediate levels of service.

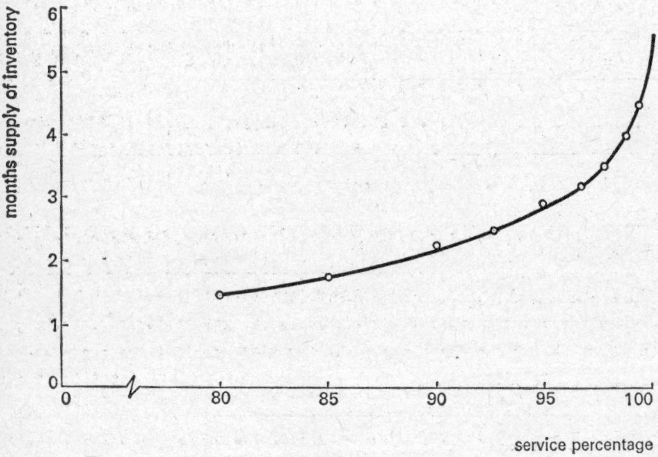

Figure 3 The price you pay to give service

If you had sufficient inventory to give 90 per cent service routinely, instead of only 80 per cent, you would gain good will, a better competitive position, lower costs of expediting and processing backorders, less loss of immediate profits, and better prospects of an increased market. Is the value to you of these benefits at least as great as the cost of supporting an additional three-quarters of a month's supply in inventory?

Ask yourself the same question for even higher levels of service. At some point, you will be forced to conclude that the value of increasing the level of routine service from inventories is not worth the cost of supporting the investment. That is where you stop and pick the corresponding safety factor.

(c) *Safety factors when there are substitutes*

Perhaps your customers' needs or desires are not entirely specific. In some instances, particularly in retail sales to consumers, the customer has a general idea of what she wants, but if one item is not in stock, she'll be satisfied with an alternative. She may never even know that you sometimes stock the other item. The manufacturer may take exception to my examples, but a grocer or druggist may see how brands of coffee and cigarettes would qualify.

In that case, you will want to select a safety factor to ensure a desired probability of selling something to the customer. That probability is

$$\text{Probability of sale} = 1 - (1 - P_1)(1 - P_2)(1 - P_3)\ldots$$
$$= 1 - (1 - P)^N,$$

where P is the probability that a specific one of N alternative items is in stock. Table 3 lists the service you should aim at on each item if you want the total level of service shown at the top of each column.

Table 3

Service Objective on Each Item when there are Alternatives

Number of satisfactory alternatives	*Desires total service for the line*				
	80%	*90%*	*95%*	*99%*	*99·9%*
1	80	90	95	99	99·9
2	55	69	78	90	97
3	41	54	63	79	90
4	...	44	53	68	82
5	...	...	45	60	75
6	...	...	...	54	68
7	...	...	...	48	63
8	...	...	...	...	58

For example, suppose that you want to have a 95 per cent chance of satisfying the customer but that you have three items that are all equally satisfactory to her. Then you would pick safety factors designed to give about 63 per cent service on each item.

Of course, this scheme works only when the customer can be satisfied with an alternative. A housewife may get a different brand of coffee because she needs coffee. A man may buy a different filter cigarette if he can't get his usual brand. Or a competent salesman, who knows his stock, may sell an appliance or automotive accessory from stock to a customer who would be bewildered if he had a complete line to choose from.

2. Safety Factors when the Backorder Cost is Known

Most authors, in discussing safety factors, are beguiled by the mathematical convenience of assuming that there is a known cost of being in backorder and derive elaborate models for selecting a safety factor that will balance the cost of carrying inventory against the expected cost of being in backorder. Frequently this assumption is merely a confusion between the *value* of avoiding a backorder and the *cost* of having a backorder, which may not be the same at all.

But let's say you run a clean, convenient, low-priced grocery store. A customer who wants a box of soap flakes will get it from you, if you have it on the shelf. If you don't have it, she will buy it next door, where the prices are higher, or will wait for her next trip downtown. In either case, you have lost the net profit on that sale and have clearly had no other appreciable expense. There's no long-term loss of good will, no paper work, no extra effort in customer relations needed. Therefore we can fairly assume a fixed, definite cost to you for each unit of demand you couldn't fill. We shall call that a backorder cost. Let's assume that it's three cents a box.

If you order soap once a day (Q = one day's supply) and if it costs 0·005 cent to carry a box in inventory for one day, then you ought to use a safety factor of 2·95.

There is a formula that gives the answer. First, form the fraction

$$1 - \frac{\begin{Bmatrix}\text{cost to carry one unit in}\\ \text{inventory for one month}\end{Bmatrix} \times \begin{Bmatrix}\text{month's supply}\\ \text{ordered at one time}\end{Bmatrix}}{\text{cost of backordering one unit}}.$$

Go back to Table 2, of the normal distribution, and find the value of $F(t)$ closest to the answer. The corresponding value of

t is the safety factor that will minimize the total cost of backorders and of carrying the inventory.

In the example of the soap flakes, we'll use days instead of months. The fraction is

$$1 - \frac{0{\cdot}005 \times 1}{3} = 0{\cdot}9983.$$

In Table 1, the closest value of $F(t)$ is 0·9984, corresponding to a safety factor $t = 2{\cdot}95$. The value of the service function is $f(2{\cdot}95) = 0{\cdot}0004$, so that if we knew the standard deviation of the forecast errors, in days' supply, we could estimate the fraction of demand filled from the shelf. Let us assume that the standard deviation is half a day's supply. Then the service will be

$$P = 1 - (0{\cdot}0004)\left(\frac{0{\cdot}5}{1}\right) = 0{\cdot}9998.$$

The service, which may not be unreasonable for a grocery store, would be 99·98 per cent of demand filled from the shelf.

If it costs a great deal to carry inventory, the safety factor given by this formula will be lower. If you order a lot at one time, there's less risk of a backorder, so the safety factor can be lower, and if it doesn't cost much to have a backorder, the safety factor will be lower.

For example, suppose it cost 0·01 cent to carry a box of soapflakes in inventory one day. Then the fraction would be 0·9967, which corresponds to a safety factor of about 2·7. We would get the same result if, alternatively, the backorder cost were only 1·5 cents per box.

3. Distribution by Value

The foregoing formulas apply in the selection of a safety factor for each stockkeeping unit and could be set up as part of the data-processing routine for your inventory control, along with the formulas for forecasting and for calculating the economical order quantities. To do so, however, would be like milling the decorative slots on a hubcap: unless you are Rolls-Royce, the value of the extra precision isn't worth the cost of the necessary effort.

While all these ideas can be reduced to simple mathematical formulas, there are several degrees of uncertainty in the process: (a) the conversion factor from mean absolute deviation to standard deviation of forecast errors in a lead time; (b) the length of the lead time; (c) how well the actual distribution of forecast errors is matched by the normal distribution; (d) the marginal value to you of one point better service, or one unit less backordered; (e) the cost of carrying inventory. We now understand very well how each factor behaves on the average for a large number of items, but it isn't practical to take the time necessary to study each of ten thousand items in forty-two locations, and even if we did their behavior might easily change the minute the study was completed.

Therefore it seems logical to use the understanding of the mathematical relationships to develop a more pragmatic approach to setting the safety factor. That approach is to pick one value of the safety factor and use it initially for all items. Some items will be out of stock a little more often than you'd like: keep track of them and increase the safety factor slightly from the initial value. Others will never be short. If the inventory doesn't come close to running out, decrease the safety factor slightly from its initial value.

(d) *Routine service with a constant safety factor*

What happens to the service rendered for different items, if the same safety factor is used across the board? In general, the routine service will be better for the slow movers than for the fast movers, if the order quantities are determined for each item or category of items. That is, the vast number of items that sell very few dollars' worth a year will usually be in stock, so you don't have to pay any attention to them. The high-dollar-volume items will be out of stock more often, but the effort you put into expediting deliveries will pay off in reducing the investment in expensive inventory.

The better service for slow movers is the result of two characteristics. First, the economical order quantity (in months' supply) varies inversely with the square root of the annual dollar volume of the item. Therefore, a low-dollar-volume item will have a relatively high months' supply as its economical order quantity. It is economical to order fast movers more often, in small months' supplies. As we saw above, the larger the order quantity, the better the service with the same safety factor.

The other characteristic is that the standard deviation of forecast errors (again in months' supply) commonly turns out to be roughly proportional to the annual dollar volume. Therefore, the ratio of the standard deviation to the order quantity increases as we move from item to item in the direction of a higher annual dollar volume.

A typical curve of routine service versus annual dollar volume is shown in Figure 4.

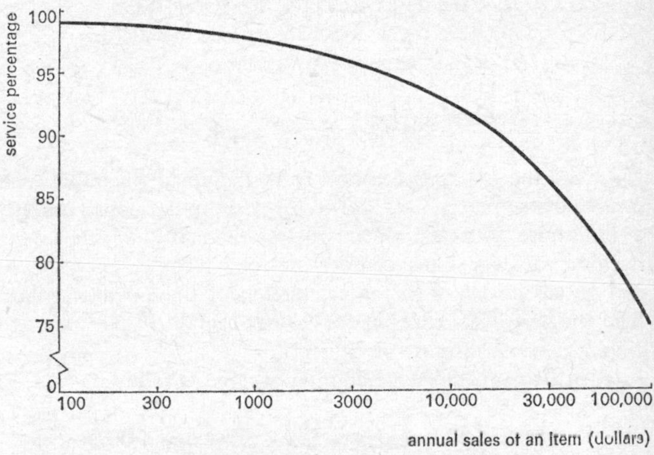

Figure 4 Per cent of demand filled from stock without delay

When you get regular replenishments in mixed truckloads, the ratio Q/M is the same for all items, and thus the serivce will also be the same.

(e) *Average service*

When each item has its own normal reorder quantity, the service routinely rendered by inventory depends, as we have seen, on the annual dollar volume of demand for the item. Since slow movers get better routine service, the expediting effort can be concentrated where it will do the most good in reducing inventory investment.

If we are to select one value for the safety factor, to use for all items, we should have some way of estimating the average service. Let us suppose that you have 50 items selling less than $100 worth a year, 20 that sell between $100 and $350, 20 that sell between $350 and $2000, 5 that sell from $2000 to $4500, and 5 that sell more than $4500 worth in a year. The average service could be computed something like this:

50% of the items × 99·0% service =	0·495
20% of the items × 98·5% service =	0·197
20% of the items × 97·5% service =	0·195
5% of the items × 95·5% service =	0·048
5% of the items × 90·0% service =	0·045
Average service	0·980

(We read the service percentage from Figure 4, using about the middle of the percentages shown for items in the dollar ranges.)

But those 50 items which sell less than $100 worth a year account for only 1 per cent of the total income. They should not be so important in the calculation of the average service. The top five items account for 24 per cent of the income and ought to be much more prominent.

Then we could calculate another average as follows:

1% of the income gets 99·0% service =	0·0099
3% of the income gets 98·5% service =	0·0296
12% of the income gets 97·5% service =	0·1170
10% of the income gets 95·5% service =	0·0955
74% of the income gets 90·0% service =	0·6660
Average service	0·9180

This system, evaluated on service weighted by dollar importance, gives an average of 91·8 per cent service.

There is a formula (see Appendix *2*) that makes it possible for you to estimate this average service directly, without quite so much calculation and graph drawing. You will need values for these factors:

$M =$ the standard deviation of forecast errors, expressed as months' supply. If you know the ratio of the standard deviation of demand per month to the average monthly sales, multiply this ratio by the length of the lead time in months.

$S =$ the average annual dollar sales rate for your product line. Just divide the total dollar sales by the number of stock-keeping units.

$\bar{Q} =$ the standard replenishment order quantity, calculated for an item with sales rate S.

Then the average service, expressed as a fraction, will be

$$\text{Average service} = 1 - 7{\cdot}31\,\frac{M}{\bar{Q}}\,f(k),$$

where $f(k)$ is the service function for the safety factor you choose. (If your immediate customers are the ultimate consumers, rather than other industries or distributors, the factor 7·31 should be replaced by 1·37.)

The safety stock, expressed as month's supply, will be just kM. The cycle stock (half an order quantity) will be a multiple of $\bar{Q}$ that depends on the distribution of your usage rates (see Appendix 1). Therefore, if you select a sequence of values for the safety factor k, you can use Worksheet 2 to calculate the total inventory in months' supply and the average service and plot a graph like Figure 5. This time, however, the inventory and the service will be for the whole product line and not just for one item at a time. Columns 1 and 2 are copied from Table 1. The service in column 4 is expressed as a fraction.

The same arguments as before apply in reaching a decision as to when the marginal investment requires costs more than the marginal improvement in service is worth. This time, I've sketched in a gray region to indicate a typical value for the service, measured in the equivalent of the cost of carrying additional investment. You cannot get the information to plot such a curve, but if you could (for this example), you'd pick a safety factor of about $k = 1{\cdot}0$, which would give 82 per cent service and require about three months' supply of inventory.

Worksheet 2

Total Investment and Average Service

Standard deviation of forecast errors $M = 1{\cdot}5$ *(Months)*

Average sales $S = \$269$ *per item per year*

Lot size for item with average sales $\bar{Q} = 6$ *(Months)*

Distribution factor $\exp(\frac{3}{8}\sigma^2) = 7{\cdot}302$ $Y = \exp(\frac{3}{8}\sigma^2)M/\bar{Q} = 1{\cdot}82$

Distribution factor $\exp(\frac{1}{8}\sigma^2) = 0{\cdot}515$ *Cycle stock* $\exp\text{-}(\frac{1}{8}\sigma^2)\bar{Q}/2 = 1{\cdot}55$

1	2	*3*	*4*	*5*	6
Safety factor	*Service function*	*Service complement*	*Service P*	*Safety stock*	*Total inventory*
k	$f(k)$	$Y \times (2)$	$1 - (3)$	kM	(5) + *Cycle stock*
0·0	0·3989	0·727	0·273	0·0	1·55
0·4	0·2304	0·419	0·581	0·6	2·15
0·8	0·1202	0·219	0·781	1·2	2·75
1·2	0·0561	0·102	0·898	1·8	3·35
1·6	0·0232	0·042	0·958	2·4	3·95
2·0	0·0085	0·015	0·985	3·0	4·55
2·4	0·0027	0·005	0·995	3·6	5·15
2·8	0·0008	0·001	0·999	4·2	5·75

Worksheet 3

Total Investment and Average Service

Standard deviation of forecast errors M = 1·5 (*months*)

Average sales S = $ *per item per year*

Lot size for items with average sales $\bar{Q}$ = 1 (*Months*) (*all items*)

Distribution factor exp $(\frac{3}{8}\sigma^2)$ = —— $Y = exp(\frac{3}{8}\sigma^2)M/\bar{Q}$ = 1·5

Distribution factor exp$(\frac{1}{8}\sigma^2)$ = —— *Cycle stock* $exp(\frac{1}{8}\sigma^2)\bar{Q}/2$ = 0·5

1	*2*	*3*	*4*	*5*	*6*
Safety factor	*Service function*	*Service complement*	*Service P*	*Safety stock*	*Total inventory*
k	*f(k)*	*Y* × (2)	*1* − (3)	*kM*	(5) + *Cycle stock*
0·0	0·3989	0·599	0·401	0·0	0·5
0·4	0·2304	0·346	0·654	0·6	1·1
0·8	0·1202	0·180	0·820	1·2	1·7
1·2	0·0561	0·084	0·916	1·8	2·3
1·6	0·0232	0·035	0·965	2·4	2·9
2·0	0·0085	0·013	0·987	3·0	3·5
2·4	0·0027	0·004	0·996	3·6	4·1
2·8	0·0008	0·001	0·999	4·2	4·7

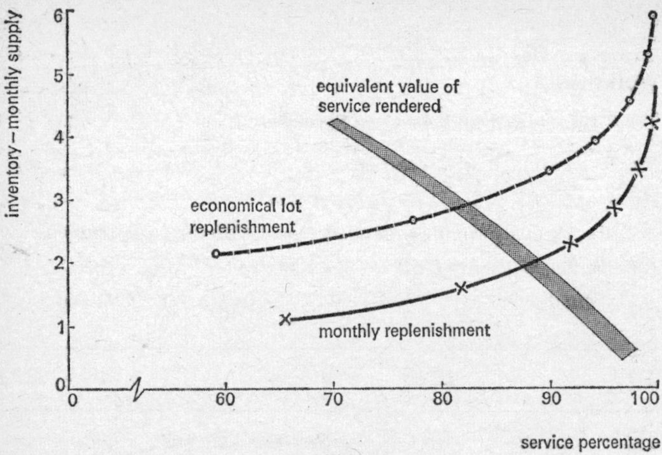

Figure 5 The total investment you need versus average service

If you review all stock levels monthly, $\bar{Q} = 1$ month. Let us assume that the standard deviation of forecast errors is still equivalent to $M = 1{\cdot}5$ months. Then the average service is the service given each item

$$P = 1 - \frac{M}{\bar{Q}} f(k) = 1 - 1{\cdot}5 f(k).$$

The service and investment in this case are illustrated in Worksheet 3. In this example, a safety factor of $k = 1{\cdot}0$ is still appropriate (Figure 5), but the service would be nearly 90 per cent with only two months' supply of inventory.

Appendix

[. . .]

1. *Average cycle stock*

Let us suppose that you compute your order quantities with a formula like (Magee, 1958)

$$Q = 12\sqrt{\left(\frac{2A}{rs}\right)} \qquad 7$$

where

Q is the quantity ordered, expressed in months' supply (and may represent either the standard lot size, or the standard interval between replenishment orders);

A is the cost of placing an order and setting up the production run. We shall assume one average value for all items in the line.

r is the inventory carrying charge, including return on investment, taxes, insurance, and obsolescence. For our purposes, the value is adjusted so that it represents the rate against selling price, rather than cost.

s is the annual sales of the item, in dollars.

The average value of the cycle stock (half the value of the standard order quantity) is

$$V_c = \frac{1}{2}\sqrt{\left(\frac{2A}{r}\right)} \text{ (average value of } \sqrt{s}\text{).} \qquad \mathbf{8}$$

If we assume that the sales rates s are lognormally distributed with standard deviation σ, the average value of $\sqrt{s}$ is, by 2 [not included here], just

$$\overline{s^{\frac{1}{2}}} = \sqrt{\bar{s}}\, e^{\frac{1}{2}(-\frac{1}{2})\frac{1}{2}\sigma^2} = \sqrt{\bar{s}}\, e^{-\frac{1}{8}\sigma^2}. \qquad \mathbf{9}$$

Therefore $$I_c = \tfrac{1}{2}e^{-\frac{1}{8}\sigma^2}\bar{Q}, \qquad \mathbf{10}$$

where $\bar{Q}$ is the standard replenishment quantity (in months' supply) calculated for an item with average sales rate. Note: If you review the stocks for all items at the same regular intervals, then $Q_i = \bar{Q}$ for all items, and $I_c = \frac{1}{2}\bar{Q}$. [. . .]

2. *Average service*

We know from the definition of the service function (section (*a*), p. 265) that the fraction of demand filled routinely from stock, without delay is

$$P = 1 - \frac{M}{Q} f(k), \qquad \mathbf{12}$$

where

M is the standard deviation of forecast errors (months' supply),

Q is the standard order quantity (months' supply),

$f(k)$ is the service function of the safety factor k.

I shall assume that your company is like every company I have worked with in that you will find that the standard deviation of forecast errors is indeed approximately proportional to the total annual sales of the item. Then, if we substitute from **7**, we find that

$$P = 1 - \frac{M}{12}\sqrt{\left(\frac{r}{2A}\right)}\sqrt{s}\,f(k). \qquad \textbf{13}$$

When we weight the service given on each item by its annual sales, the average service is

$$\bar{P} = 1 - \frac{M}{12}\sqrt{\left(\frac{r}{2A}\right)}\frac{f(k)\overline{s^{3/2}}}{\bar{s}}$$

$$= 1 - e^{\frac{3}{8}\sigma^2}\frac{M}{\bar{Q}}f(k) \qquad \textbf{14a}$$

for economic lot-replenishment, or

$$\bar{P} = 1 - \frac{M}{\bar{Q}}f(k) \qquad \textbf{14b}$$

for replenishment on a regular schedule.

References

BROWN, R. G. (1957), 'Economical safety stocks', Fifth Annual Meeting of ORSA, Philadelphia.

MAGEE, J. F. (1958), *Production Planning and Inventory Control*, McGraw-Hill, ch. 4.

Part Four **Systems Planning**

There is a 'how-to-do-it' mythology, and it goes something like this: We should start with system's planning and we should end with it. The ending system should be larger, more comprehensive and inclusive than the beginning one. The ending system constitutes the beginning of the next round of system's planning.

Can we really aggregate in this way? The answer is that we can *not*! We can learn in this way, moving gradually from the parts to the whole, but we cannot design the large system in this manner. There are irreversible sub-optimizations that will occur as we aggregate 'optimal' subsystems. (In effect, Σ (best subsystems) $<$ best Σ (subsystems).) So we have to discard the incrementalism of Couéism ('every day in every way I get better and better') and start out by asking, 'What makes a man good?' The mythology of small steps must go. We have to work on small systems and the large ones simultaneously. The articles in this section provide excellent exemplification.

18 H. Goode

A Decision Model for a Fourth-Level Model in the Boulding Sense

H. Goode, 'A decision model for a fourth-level model in the Boulding sense', in D. P. Eckman (ed.), *Systems: Research and Design Proceedings of the First Systems Symposium at Case Institute of Technology*, Wiley, 1961, pp. 105–17.

Modeling systems leads to better understanding from both the practical standpoint of designing large-scale systems and the theoretical objective of making progress in unifying a system theory. I shall begin by adopting a hierarchy of system levels due to Boulding (1956) which is useful in pinpointing my system level of interest. The latter is the 'cell', Boulding's fourth level, an open system capable of maintaining its integrity in the face of a throughput of material, energy, and/or information. My special interest is in decision cells, and to make my ideas concrete I shall examine two practical systems, underscoring the points of decision and isolatable decision cells. To narrow the problem, the decisions made in each of these cells are then classified and are found to be one of three types, of which I select one. These types are estimation, multiple alternative, and go, no-go decisions, and the third class is the one chosen for discussion. Many of the decisions are in this class, which is representative of the decision process[1] and at the same time is simple enough to expedite exposition of a model.

To delineate the parts of the 'cell' model for the decision class chosen, I shall give a brief historical sketch of the development of the theory of go, no-go decisions, leading up to a modern statement of this type of decision in terms of three elements: a priori, productive probabilities, and values. With the parts of the decision made firm, a model is depicted which I believe adequately represents go, no-go decision making at the cell level. We are then in a position to see what light the model can throw on some of the aspects of system design which have come to

1. In a sense, all of these problems are the same, the mathematics being of a similar type but more complicated in the first two.

notice, such as the standardization of information, the centralization of the system, etc. Some of these aspects, in the light of the model, can be sharpened to principles. Based on this discussion, conclusions can be drawn concerning the usefulness of such models and extensions can be suggested.

The Boulding Classification

In his article on general systems theory, Boulding sets forth a suggested hierarchical classification of systems as shown in Table 1.

Table 1
Hierarchy of Systems

Key word	*Characteristic*	*Example of system at level*
1. Framework	Static structure	Geography of earth, anatomy of cell
2. Clockwork	Predetermined motions	Steam engine, solar system
3. Thermostat	Transmission of information	Furnace homeostasis
4. Cell	Open system throughput of material and energy, system maintained in the face of throughput	Flame, river, cell
5. Plant	Division of labor among parts but not at the sensory level	Any plant
6. Animal	Self-awareness, specialized receptors, increased intake of information to form image (different from the information)	Any animal
7. Human	Self-consciousness, knows that it knows, language and symbols	Human being
8. Social organizations	'Role' and communication, history, messages, music, art	Human societies
9. Transcendental systems	Unknowables	Unknowables

It is clear that our understanding decreases with system level. As Boulding points out, most of our knowledge is at the first, second, and third levels. We are just beginning to grasp understanding at the fourth level, and my discussion is pointed toward modeling decisions at this level.

Practical Systems

Consider the air defense system. In one sense it is composed of a series of decisions which are of relatively simple types, each of which is performed by an elementary 'cell'-type mechanism. Beginning at the sensory end, the system depends upon an unattended radar device which scans through a quantified set of regions on an oscilloscope and decides, on the basis of the number of energy returns above a fixed level, chosen in advance, whether a 'target' is present at any particular scope region (corresponding to some particular space region). This decision is of the go, no-go type. It is passed on, if positive, to another position of the system which begins the establishment of a 'track' on the detected target. This decision is of the estimation type, and it estimates numbers, course, and speed, which characterize the horizontal track. The system continues, with the analysis in Table 2 yielded by examining each step of the system.

We note first that the decisions are classifiable in a small number of classes, and second, that the go, no-go type occurs quite frequently.

A second system which may be examined on this basis is the path of a letter through an automatic post-office system. The letter arrives at the post office for a first decision concerning whether it is package or letter, then faced (if letter) into a required orientation which is one of four, then directed to local or out-of-town mail, then read for coding of address, then interpreted, etc. The system analysed yields the following breakdown:

Culling	Go, no-go
Facing	Multiple choice
Local or outgoing	Go, no-go
Coding of address	Multiple choice
Interpretation	Multiple choice
Sort	Multiple choice

Table 2
Air Defense System

Class of function	*Function*	*Class of decision*
Front end analysis	Detection	Go, no-go
	Track determination	Estimation
	Height determination	Estimation
	Split targets	Go, no-go
	Merge targets	Go, no-go
	Identification	Go, no-go
Raid decision	Speed	Estimation
	Direction	Estimation
	Strength	Estimation
	Attack	Go, no-go
Attack	With what	Multiple choice
	How many	Multiple choice
	From which base	Multiple choice
	When	Estimation
	Kill or no-kill	Go, no-go
Weapon direction	Acquisition	Go, no-go
	Track	Estimation
In missile	Seeker acquisition	Go, no-go
	Warhead detonation	Estimation.

Many of the multiple choices are automatically performed by a go, no-go mechanism at each of many possible choices. Similar breakdowns are possible for many systems. Each exhibits a 'cell-like' structure and in the cells are decisions of a small number of class types, an important one always being go, no-go (or its generalization, multiple choice).

Go and No-Go Decisions

We seek now to understand the elements of a go, no-go decision. The early investigators of decision theory, Laplace and Poisson, tried to make decision statements in terms of probability values with no particular worry about the logical basis of their statements. Laplace introduced a formula for a go, no-go decision

$$\frac{r+1}{n+2}, \qquad 1$$

which stated that the probability of success on the next trial of an experiment was

$$p(s) = \frac{r+1}{n+2},$$

where r is the number of past successes and n the number of trials thus far. The probability of the sun rising tomorrow is therefore pretty close to 1! To apply this formula to go, no-go one noted the probability as of the last observation and bet according to ones' intuitive demand for odds.

Bayes connected the state of knowledge before an experiment with the knowledge from the experiment to decide between go and no-go by writing a formula for the probabilities of the two possibilities after the experiment in terms of the 'probability before the experiment' and the 'probabilities of experimental outcome' in case each alternative might be true:

$$P_x(S) = \frac{P(S)\,l(x)}{P(S)\,l(x) + P(N)}. \qquad 2$$

Thus, if our alternatives are designated hypotheses (null and special, i.e. no-go and go) and if each has *a priori* (before the experiment) probability $P(N)$ and $P(S)$, and if an observation x occurs from the experiment with $P_N(x)$ when N is true and $P_S(x)$ when S is true, then Bayes wrote

$$P_x(S) = \frac{P(S)\,P_S(x)}{P(S)\,P_S(x) + P(N)\,P_N(x)},$$

where $P_x(S)$ is the *a posteriori* probability that S is true if x is observed. Similarly, for $P_x(N)$ we can rewrite the formula

$$P_x(S) = \frac{l_a(H)\,l(x)}{l_a(H)\,l(x) + 1},$$

where $$l_a(H) = \frac{P(S)}{P(N)} \quad \text{and} \quad l(x) = \frac{P_S(x)}{P_N(x)}$$

and the latter is called the likelihood ratio, already putting in evidence a quantity which turns out to be more and more important. Again decision is made according to one's fancy concerning the required probabilities of being right.

For a little while this seemed to solve the problem of going backward from experiment to hypotheses using the knowledge to date and the knowledge gathered from the experiment. But it bothered many that when $l_a(H)$ was not known people began to put unity in its place, announcing a principle of indifference, or ignorance, or what have you. Bayes' theorem fell into disuse.

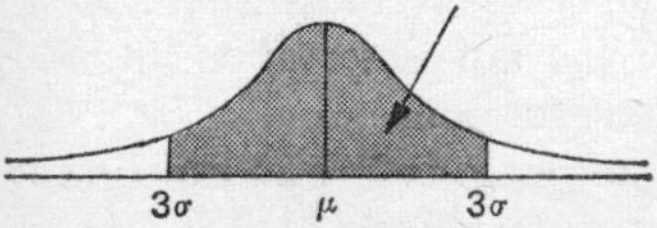

Figure 1 Classical theory

Over the years, large number theory developed with the distinction between the null and special hypothesis becoming blurred (Figure 1). A rather loose use of probability gave rise to the statement that the 'population mean lies within $\pm 3\sigma$ on either side of the sample mean' with no attention paid to either *a priori* knowledge or the alternatives presenting themselves or the values of decisions.

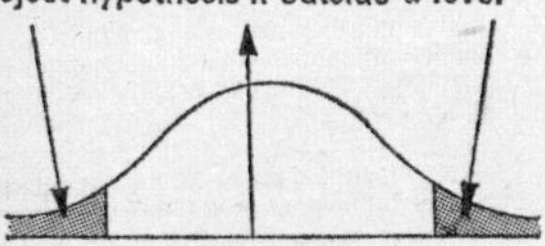

Figure 2 Fisher's theory

Fisher, dissatisfied with this state, devised the 'test of significance' which, although it introduced neither prior knowledge nor alternative hypotheses, did make a precise statement on a go, no-go basis (Figure 2). If you choose a null, Fisher said, you may adopt a policy which will fix the probability of rejecting it in error.

Later Neyman and Pearson pointed out that one did not first accept or reject a hypothesis but almost always had another alternative in mind which would be rejected or accepted as a result of the decision (Figure 3). They introduced the notions of type-I error (rejection of null when true) and type-II error (rejection of special when true). The likelihood ratio turned out to be the key to obtaining the smallest β (type-II error rate) for

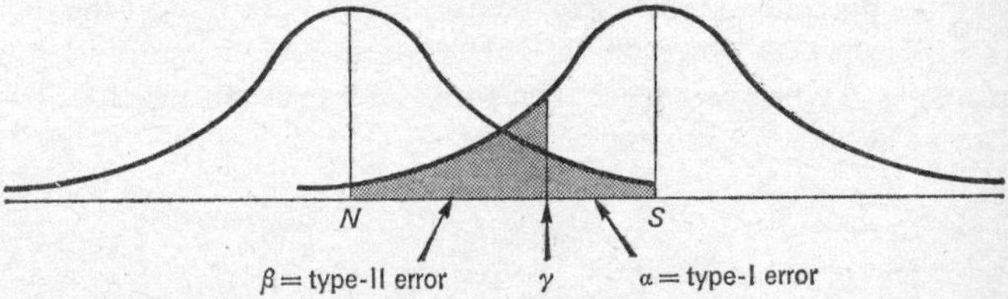

fix α: then β is determined by distributions and distance between means

Figure 3 Neyman and Pearson theory

a given or fixed α (type-I error rate). This development brought the alternative hypothesis back into the picture but still left out of consideration two important desiderata: the knowledge to date and the costs of making errors.

A. Wald took into consideration the latter with statistical decision theory, which introduced a cost function to be applied to the type-I and type-II error rates:

Hypothesis

Guess	N	S
C	V_{NC}	V_{SC}
A	V_{NA}	V_{SA}

Choice of γ:

$$\frac{P_S(x)}{P_N(x)} = l(x) \geqslant \frac{P(N)}{P(S)} \frac{V_{NC} - V_{NA}}{V_{SA} - V_{SC}} = \gamma \qquad 3$$

The theory played against nature as though she were an opponent and therefore ascribed no *a priori* knowledge to the experiment.

Recently, a signal detection theory has taken the last step toward bringing all elements back into the fold. By applying

Bayes' theorem **2** to Wald's decision function **3** and Neyman's and Pearson's alternative hypothesis (Figure 3), we have the following theory of the go, no-go decision: Given two hypotheses (alternatives), Null (N) and Special (S), whose *a priori* probabilities of occurrence are $P(N)$ and $P(S)$, a decision A implying belief in S and a decision C implying belief in N; the costs of correct statements and error being V_{NC}, V_{SA}, V_{NA}, V_{SC}; the distribution of an observed variable x being $P_N(x)$ when N is true and $P_S(x)$ when S is true; then to make a decision which maximizes the expected value, compute

$$l(x) = \frac{P_S(x)}{P_N(x)}. \qquad \textbf{4}$$

If $$l(x) \geqslant \frac{P(N)}{P(S)} \frac{V_{NC} - V_{NA}}{V_{SA} - V_{SC}}$$

say that S is true. Otherwise, say N is true. The cutoff point γ for decisioning is chosen so as to *maximize the expected value* based on knowledge of costs, *a priori* probabilities, and productive probabilities. More detailed consideration shows that even if the expected value is not the criterion, $l(x)$ is the proper indicator, but a new criterion formula must be computed.

Example from Practical System

Consider the first 'cell' of the air defense system, the unattended radar. At any region on the scope, a decision is to be made that just noise is present, or signal (plus noise). Suppose that the *a priori* probability of a signal is $P(S)$ (of course, a function of many things including hot or cold war conditions). When a signal is present, the returning radar energy will be collected in some indicator reading x in value: x will be distributed in a certain fashion, $P_S(x)$. When only noise is present, x will be distributed differently. If, further, the cost of saying signal when there is a signal, noise when there is a noise, false alarm, and the target miss are evaluated, then **4** determines the automatic cutoff point for setting the radar. Obviously, conceptual insight rather than explanatory prediction has been achieved through this equation since many of the quantities are unknown.

An Open System or Cell Model

The value of the foregoing discussion is that it leads to isolation of the important parts of an elementary decision. A system operating to make a decision may be envisaged as in Figure 4. This diagram was originally conceived by Quastler, modified by W. P. Tanner, and again modified by the present author.

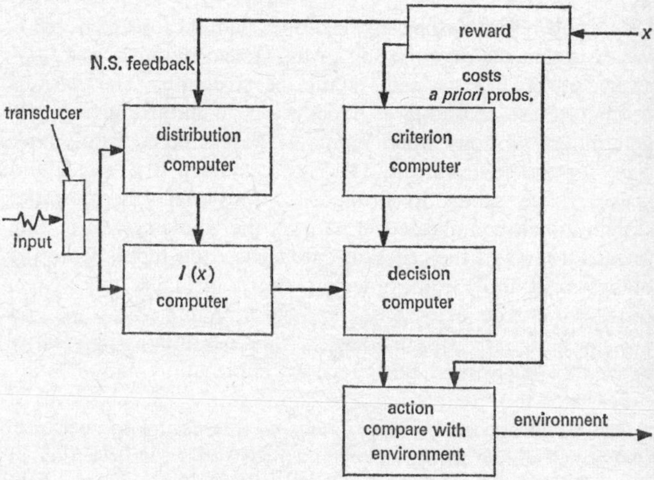

Figure 4 A decision system

Energy (or material) is passed into the system through a transducer to provide the standardized format of the 'cell'. From all the irrelevant information present (which may involve millions of bits), what is wanted is the likelihood ratio, i.e. the ratio of probabilities that this phenomenon will be observed when *S* is in operation to that of observing it when *N* is in operation. However, this value cannot be computed without keeping records of these distributions for introduction to the $l(x)$ computer. Moreover, the $l(x)$ value has meaning only as it can be compared with a criterion to make a decision. The criterion depends on what we have learned to date (the *a priori* probabilities) and the costs of making correct statements and errors. This information comes from comparing decisions made with rewards achieved.[2] The reward box, its measure and reactions, comes from some external system and for this reason we are dealing with an elementary cell rather than with specialization.

Application of the Model

Let us now return to the classification system set forth in Table 1. We note that the organization of the system by boxes is a first-level 'framework' system, static in structure. The various computers are second-level 'clockwork' in nature, giving predetermined motions when inputs of values come from other boxes or the environment. The feedback from the reward and action boxes serves to produce a third-level 'thermostatic' system (albeit complicated in nature), the whole system pulling toward a level of the criterion cutoff based on inputs from the outer world and the meaning of rewards. Finally, this system processes energy or material and information and passes on decisions, yielding an open system which maintains its identity in the face of a throughput of energy, information, and material.

It is not too great a jump to suppose that the next level of system is composed by sets of cells of this nature to specialize functions relative to the development of reward. Interaction of rewards might give a 'plant'-type system.

Principle – the Standardization of Information

I should like to emphasize again that all the information from external sources necessary for any given decision is in the likelihood ratio. Now experience shows that the designer of any large system finds it desirable to standardize the form of the information flow in the system early in the single thread development.

2. Insight into the meaning of the principle of indifference comes from asking what values one would ascribe to such a device initially when no experience has been gained. After the first random decision (yes and no equally – which says a coin is tossed to say whether $l(x) >$ criterion), $P(N) = P(S)$ and all Vs equal seems the only way. $l(x) > 1$ is the criterion.

The telephone system turns the number which used to be embedded in lots of words into a digital code which then passes essential information around the system. The air defense system turns the statement about a target into a code. The post office is beginning to turn its addresses into code early in the system. But our model tells us something about how to standardize. The likelihood ratio contains all the information. Therefore, it is the number to be passed on if no decision is to be made. Moreover, to carry other stuff along is wasteful of memory and computer capacity as well as confusing. To avoid this waste, standardization must occur early in the system, in accordance with a practical system design notion.

To Decide or Not to Decide

But should a decision be made? Two kinds of deferments in decision making are recognized as frequently required in practical system design. One refers to the requirement for more information, the other to a requirement for decision at a higher level. These two points will be separately discussed under Deferred Decision Theory and Centralization and Decentralization. Here I am concerned with only the information aspect of decision making. Executing the decision destroys information.

To see that this is true, consider the air defense example. If two radars are observing and both have signals slightly below cutoff, the information will be discarded if both decide *at the radar* whether a target exists. If both had held likelihood ratios, the product of the ratios would be above cutoff values leading to a higher probability of correct decision. Therefore, except as modified by the time requirements of the environment and the memory requirements, decisions should be held off.

Deferred Decision Theory

It is possible to extend the model of a decision to include three possible actions instead of two: i.e. instead of deciding that one hypothesis or the other holds, a third alternative allowing deferral of decision pending more information is allowed. This theory, under a maximum expectation criterion, makes a decision now whenever the expected value for waiting is less than that available for making a decision. The calculation of cutoff points is at present impractical, but again involves the likelihood ratio, this time leading to *two* cutoff points on the $l(x)$ axis: below 1, accept Null; above 2 accept Special; between the two hold onto the information. In the simple case where the curve of probability of a correct decision has a non-increasing slope, a simple solution is possible.

Memory Considerations

But holding onto information demands memory capacity. I believe that limitations of memory and the ability to search efficiently are the most serious limiting factors in our development of systems of great sophistication. We shall experience a revolution in automation when the memory problem has been adequately solved. Again, consider the radar example. For a scope cut into 10,000 regions we would have to remember the likelihood ratios for each region, and *transmit them*, if we deferred decision. For this reason, we do not. But we are going through a metamorphosis in memory-handling capability and 10^5–10^7 bits per cubic inch is predicted. Estimates of human capacity are 10^8—10^{11} bits per cubic inch. We begin at last to get within sighting distance. Our system model shows the extent of the memory which we must seek.

Centralization and Decentralization

If we examine the various forces leading to centralization of a system, or the opposite, we see that on the side of centralization is an integrated decision using all the information available (all likelihood ratios), plus the possibility of co-ordination of the criterion center to yield reward at a level above the 'cell' unit. In fact, if the cell only is to be optimized, we may see at once that centralization yields nothing. Further, central actions may be highly undesirable from the standpoint of the particular 'cell'. Thus, using *all* information and acting for the good of the larger system are the payoffs for centralizing.

On the other hand, memory requirements and computing

requirements yield attenuation, high error rates, and long delay times in response, as well as congestion at the clearance center. We draw several conclusions. We must decentralize for the practical design reasons of inability to handle at the center. But changing technology demands a repeated examination of the point of decentralization. Finally, the less the coupling between cells for the survival of the system (or its reward), the less the gain from centralization.

This latter point explains the decentralization of functions in living organisms to a level sufficient to the part of the system at stake. In all probability, the cell in the body handles all of its own homeostasis at its own level, and passes on information pertinent to survival of the group. The implications for practical system design are evident. Too frequently, system engineers are biased toward centralization or decentralization with no rational basis. The model demands examination of (a) the reward system, (b) the memory and computing capacity, and (c) the attenuation, error, and delay rates.

Subunit Size

We draw a corollary conclusion concerning the size of a subsystem. Are its decisions pertinent to itself or close-coupled to a larger piece? If the former, allow to operate as a subsystem; if the latter, do not allow autonomy. This yields a complicated structure if several functions are being carried on, but the design principle is clear.

Adaptive Systems

We hear much today about adaptive, or self-adaptive systems. We may use our model to examine the notion. Adaptive systems lie somewhere between the fourth and fifth level in the Boulding classification. At the lower levels of system performance, material and energy are absorbed (feeding), the process yielding survival of the system (and, in some systems, reproduction) and subsequent disposal of waste and/or conversion of energy. On the next higher level of sophistication, the system processes information, sensing it, communicating it, making decisions (reasoning?), and effecting actions. That is the fourth level I have been describing. It may, however, carry out this process according to an accepted script with no modification of the process, or, on the other hand, the information, material, and energy entering the system may change its configuration to perform each of the lower level functions, in which case the system has 'adapted' or learned. But this adaptation may be at a very low level, the result of the realization of one of several previously foreseen states of the environment, as, e.g. changing distribution values (according to a rule provided), changing criterion values (not criteria), etc., or it may adapt to a completely unforeseen environment by changing criteria and functional form. Still it has had no changes in goal or objectives. For this type of adaptation it must have its reward box[3] modified, which implies some change at a higher system level. We are now in the order of the fifth and sixth levels. We have thus stated our notion of adaptation in terms of the 'cell' model.

Conclusions

From this discussion I conclude that we may begin to interpret our 'practical' system design considerations in terms of the model, and occasionally extend the meaningful content of these considerations. Often we may correct the notion, or on occasion discard it.

Clearly, the model needs extension in two directions: first, by combination with notions from other disciplines than statistical hypothesis testing, such as information theory and game theory; and second, by beginning its extension to the upper levels in the system hierarchy – specialization of functions and awareness.

3. The reward box is a stored knowledge of what is 'good' and 'bad' for the system.

Reference

BOULDING, K. E. (1956), 'General systems theory – the skeleton of science', *Management Science*, vol. 2., p. 197.

19 J. A. Alcalay and E. S. Buffa

A Proposal for a General Model of a Production System

J. A. Alcalay and E. S. Buffa, *A Proposal for a General Model of a Production System*, Wiley, 1966, pp. 36–52.

The Production System in the Organization

The production system must be recognized as being a subsystem of the total organization system which in turn is a subsystem of the 'universe system'. In Figure 1 and elsewhere in this section the systems external to the organization are referred to as the 'environment systems'.

The implicit goal of any total organization system is to maintain stability under as wide a range of time variations of its interactions with the environment systems as is possible while advancing by optimal degrees toward a prescribed goal. The concept of stability here is identical to the one developed by systems engineers.

The following three functions of management are of primary importance and can be approached by systems engineering methods.

1. *The analysis function.* The dynamic characteristics of each organization subsystem are studied, that is, the purpose is to discover how a subsystem acts upon a set of input variables to obtain a set of output variables.

2. *The design function.* Knowing the generalized transfer function of all subsystems, the design function characteristic parameter may be changed with the purpose of increasing the organization stability and performance index. This change can be achieved either (a) by improving the quality of the 'component elements' of the system (e.g. better personnel, better machines, better work methods, better quality of organization information, material and energy transfer channels, etc.), or (b) by introducing compensating subsystems (e.g. introducing individual control subloops or additional measurement subloops, creating 'expediting' groups, functional and service departmentation) or by removing

superfluous subsystems and changing the information, material decision flow patterns. (Note that the concept 'information' here is not used in the sense defined by Shannon and other information theory specialists, but rather in its more common usage meaning conveyed communication with intelligible meaning.)

3. *The control function.* The first two functions may be (and in this paper are) considered to be part of the control function. This function will be further discussed below in conjunction with its related subsystems. It is mentioned here because of its importance. This function consists in varying some or all of the independent variables of each subsystem within some constraint limitations and on the basis of knowledge of the dynamics of each subsystem, measured and gathered data, and in the direction determined by policy and performance criteria. The control function is executed by control subsystems. Thus the functions of analysis and design of these subsystems as described above apply equally well.

Mathematical Statement of the Problem

Since the three functions discussed previously apply equally well to any system or any properly defined subsystem, the statement of the problem will apply equally well to any system and in particular a production system that is a subsystem of an organization system.

Consider the optimization of a general production system with many inputs and outputs. An input is defined as any quantity of interest that has an effect on the system and is controllable. Output is defined as any quantity of interest that is completely determined by the values of the input variables and the dynamic characteristics of the system. It is convenient to think of the system as a mathematical operator M which operates on the set of inputs (u_i) to yield the set of outputs (v_i).[1] Note that the operator M is generally multidimensional and highly non-linear.

The system is represented symbolically by equation **1**

$$M[\{u_i(t)\}] = \{v_i(t)\}. \qquad \mathbf{1}$$

This equation is read: The operator M operates on the set of input variables $u_i(t)$ to yield the set of output variables $v_i(t)$. The functional notation (t) indicates time dependence, where in our

1. A glossary of symbols used is included at the end of the chapter.

case the time dependence is caused by quasirandom fluctuations in the input variables. The system has an objective function associated with its operation. In its most general form this objective function includes all the important variables related to the system, as well as variables and characteristics of systems of the environment, denoted below by Q. When considering the production system alone, Q includes the characteristics and variables of the other organization subsystems and organization environment systems. The profit or objective function is most often determined on the basis of economic and management considerations. In production systems it is very often simply the financial profit function. This is not generally true, however. For instance, it may include such considerations as decisions to 'stay in the business' even by acquiring financial loss with the purpose of accumulating knowhow or eliminating competition. The profit is symbolized by the identity

$$B = B(\{u_i\}, \{v_i\}, Q). \qquad 2$$

In general not all operating levels of the system are allowed, since constraints are quite often imposed either by the system itself or by characteristics of environment related systems. For example, the rate of output may be limited by the output limitation of a machine (system imposed constraint) or by the Fair Labor Law (environment constraint). Limits of this type are called constraints and are represented by

$$C_j(\{u_i\}, \{v_i\}) < 0, \qquad 3$$

where C_j is the symbolic notation for the jth kind of constraint.

With the system so defined the generalized optimization goal can be stated in detail. Locate the set of input variables x_i which for given conditions Q maximizes the profit function and also satisfies the constraints: in symbolic form

$$\operatorname*{Max}_{x_i} B(u_i, v_i. Q), \qquad 4$$

subject to equations **1** and **3**.

The three managerial functions as stated in the previous section are related to the above mathematical statement as follows. In the analysis function we attempt to determine the format of the operator M and its characteristic parameters. In the design function we attempt to modify the format of the M operator and to change its characteristic parameters with the purpose of widening the range of possible variations of the variables between the constraint limitations and hence changing the function P. In other words in the design function we attempt to find an M such that equation **4** will represent a maximum of all possible maximums for given conditions of Q.

The control function in a restricted sense performs the relation expressed by **4** but in a more general and extended sense also achieves the purpose of the design function.

The present paper restricts itself to the problem of determining and formulating a conceptual framework for a mathematical model of the operator M for a generalized production system.

Defining the Boundaries of the Production System

The definition of a production system in this paper follows closely, but not entirely, the definition given by Buffa (1961), viz., 'Production is the process by which goods and services are created', and is adopted in the strict sense here. Some of the functions included in that definition of a production system will be relegated, however, for reasons stated below, to separate subsystems. Notably all functions dealing with processing of information of any kind, decision, planning, design and control functions will pertain to other subsystems.

The reason is the possibility of a more systematic mathematical analysis when subsystems are defined as performing *single* defined functions.

Thus a production system's *single* function is the 'creation' of goods and services whose primary destination is a consumer. For the proper discharging of this function, the production system requires inputs from other subsystems of the organization such as service inputs (e.g. maintenance, supervision, product and plant layout design, research and development) and control inputs (e.g. measurement, 'book-keeping' or data processing, planning, control, order and sales information processing, forecasting, etc.).

Forrester (1961) tends to define the subsystems more on a geographic location basis. For instance in analysing the production–distribution system he defines as subsystems (a) the factory, (b) the factory warehouse, (c) the distributors, and (d) the retailers. One finds that many of the transfer functions are identical. Each of his subsystems is characterized to a great extent by a pure delay or transportation lag or as he calls it 'pipeline' delay functions.

The approach in this paper is based on a functional definition of subsystems. One subsystem will describe a single function or a component of a function, which may be executed by many persons, machines in different geographic locations. The physical entities and their geographic locations instrumental in the execution of the system function are considered only inasmuch as they affect the magnitude of the parameters of the system function.

The subsystems and their functions in an organization (Figure 1) are:

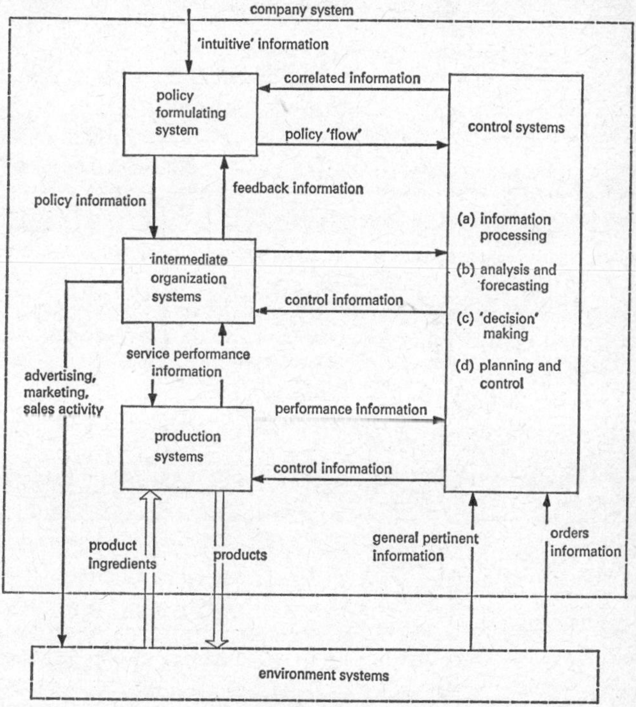

Figure 1 Interaction among production subsystem, other organizational subsystems, and environment systems

1. *Policy-formulating system.* The function of this system is to adapt basic organization policies to correlated information reflecting present and forecasted future conditions. This function could conceivably have been incorporated in the general control function. In both systems the basic function is gathering and data processing of information with the purpose of control and planning. The only reason the policy-formulating system is defined separately is that it has a line-authority organization characteristic which is not necessary in the control system. As a

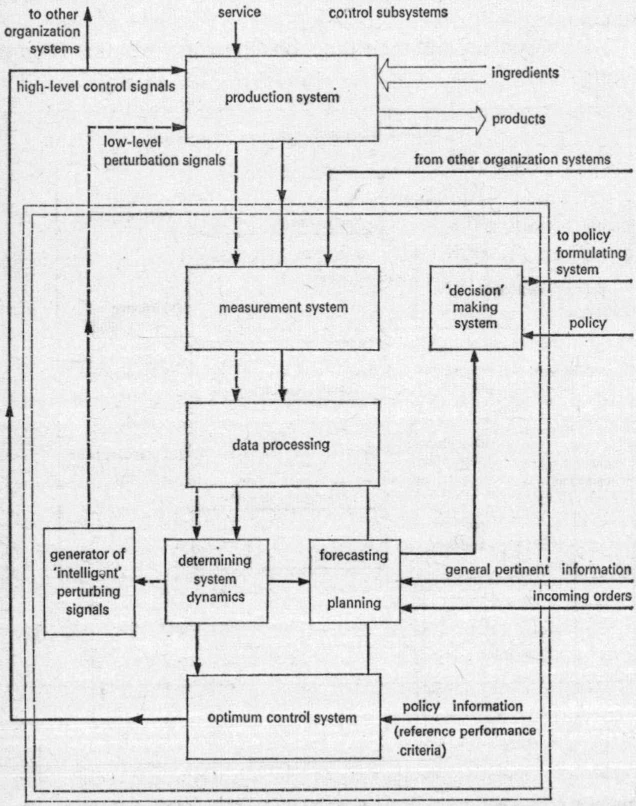

Figure 2 Production system in relation to control subsystems

by-product of this distinction one can also cite the fact that the human behavioristic characteristics can conceivably be completely absent from the control systems.

2. *Control system.* The basic function of the control system is information transformation. A possible subdivision of this function into component functions (and therefore subsystems) is illustrated in Figure 2. It is conceivable that the transformation of information can be completely handled by machines and computers. The control system has a purely advisory or staff authority organization characteristic.

3. *Intermediate organization systems.* The emphasis in this paper being analysis of the production system, the intermediate organization systems have not been clearly subdivided into single function subsystems. If, contrary to accepted terminology in management theory, supervision and transfer of decisions are incorporated in a generalized meaning of service (the 'service' of supervising or the 'service' of delegating authority and transferring decisions), the single function of the intermediate organization systems could be formulated as follows: To provide the necessary service to other subsystems of the organization or to some subsystems of the environment which directly affect the organization subsystems. Some of these services (following more or less conventional service departmentation criteria) directly related to the production system are illustrated in Figure 3. The same figure shows some of the data-collection (measurement) locations and many of the control-element locations. These, of course, following the present scheme, are part of the control system.

4. *The production system.* Its function was stated above and it will be the exclusive concern of the following sections.

The Production System

The functions and the boundaries of the production system were defined in the previous section. In this section its subsystems and component functions are defined. A product j tree represents the

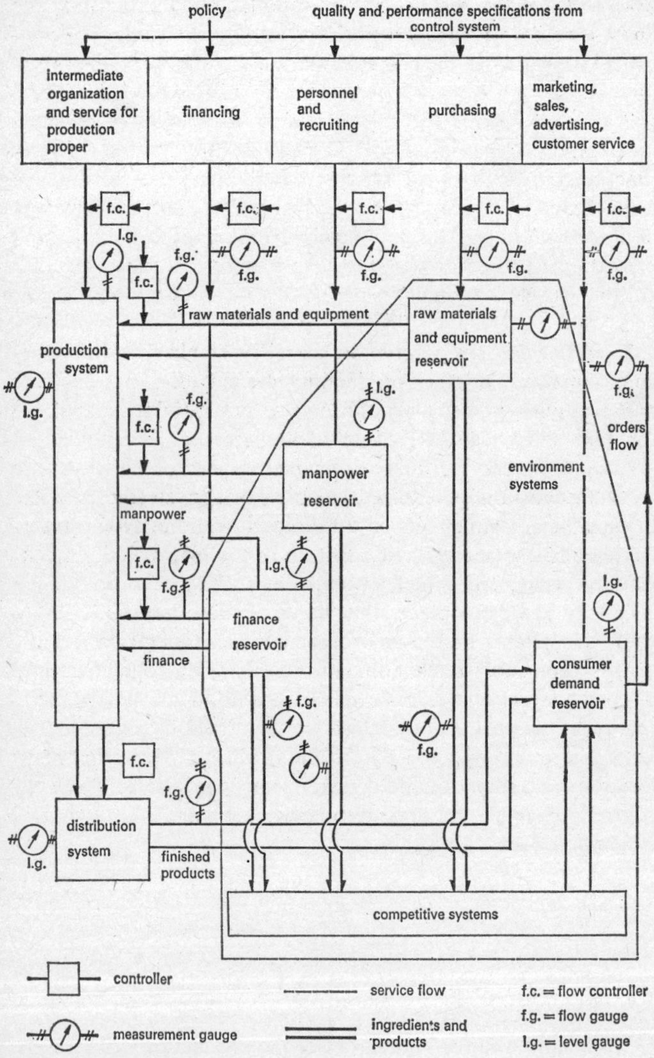

Figure 3 Relation among production system, service systems, and some environment systems

set of operations and processes necessary in producing a *single* final consumer product of unique specifications denoted by *j*. A product *j* tree may use intermediate products originating from other trees and it may produce intermediate products destined for use by other trees. The only restriction is that a single product whose destination is the consumer is produced by a tree. A product *j* tree may also produce products which are generally undesirable. These products are called waste products and their destination is the environment.

A *j* tree can be visualized as consisting of a number *n* of sequential stages. For the present analysis a stage k_j is defined as a tree *j* subsystem in which a subset of operations and processes on one or more types of products is considered. Furthermore a stage has at least one input or output product stream not originating from or directed to a stage sequentially adjacent, except if the product is a waste product.

1. The *distribution system* is considered to be a stage of the product tree (though of a special restricted nature). If the last transformation stage of a product tree is denoted by $n - 2$, the distribution system will always be $n - 1$.

2. The *consumer system* is not to be confused with the environment consumer demand reservoir system as shown in Figure 3.

3. The *consumer subsystem* is always considered to be the last stage, stage *n*, of a product tree. It is also of a special restricted nature. It selects and classifies the project *j* fed to it into an adopted product stream and a rejected stream returning to the transformation system for rework. It does not have any inventory level and its manpower, equipment, energy, and service subsystems are nonexistent.

Subsystem Classification

There exists a logical pattern in the classification and definition of the production subsystems. This classification is presented in a matrix format (Figure 4).

Each column in the matrix represents a set of subsystems of each one of the subsystems represented by the column to its left. Thus the single product tree *j* column (set 1) represents the set of subsystems of the transformation system. Each subsystem of this set is characterized by a single consumer product j, $1 \leqslant j \leqslant m$, where m is the number of different consumer products created. Similarly, the stage k_j column represents the set of stage subsystems k_j, $1 \leqslant k_j \leqslant n_j$ to each product $_j$ tree system.

set →

0	1	2	class ↓
transformation system	production tree $1 \leqslant j \leqslant m$	operation stage $1 \leqslant kj \leqslant n_j$	
material i			0
manpower i			1
equipment i			2
energy i			3
service i			4

Figure 4 Matrix defining production subsystems

Any one of the columnar subsystems can in turn be subdivided into the five classes of subsystems. Thus one can consider the M_{ijk} subsystems, each one of which is defined as the manpower system describing the dynamic behavior of manpower of skill specification i working at stage k of the j product tree.

It becomes apparent that the number of elemental subsystems represented by a single square in the third column of the subsystem matrix is:

$$\sum_{j=1}^{m} \sum_{k=1}^{n_j} \sum_{i=1}^{S_{kj}} i \cdot k \cdot j,$$

where S is the number of skill specifications, necessary in stage k_j.

The functional format, however, of the equations describing the subsystem behavior is the same for each set or each class of subsystems. There may exist differences in the values assumed by the parameters for each elemental subsystem. This, however, as far as the computer performing the control function is concerned, is only of data storage of memory availability importance. This important observation is the logical cornerstone for the justification for the system structure analysis approach developed and presented above.

Definitions and Nomenclature

A *stream* is an entity in transition between two defined and bounded systems. The rate of flow of the entity in the stream is symbolized by:

1. A lower-case Latin character characterizing the stream.
2. Three subscript numbers.

The three Latin characters used in characterizing streams in transition between stage kj subsystems are:

p: denoting a stream directed toward the end product, called a 'product stream'.

r: denoting a stream which has to be reworked because of quality deficiency, flowing in a direction opposite from the end product.

w: denoting a nonrecoverable stream also directed in a direction opposite from the end product, but whose destination is always the environment.

The r streams will also be referred to as 'reject streams'. The w streams will be referred to as 'waste streams'. The 'decision' whether a unit is recoverable is a component signal of the control information input originating from the control system. This signal sets the 'go-no go' set point of a 'filter control box' which is also part of the control system.

The subscript number i denotes a stream of general specification i (e.g. $\frac{1}{4}$ inch bolts may be symbolized by $i = 4$; $\frac{1}{2}$ inch bolts by $i = 2$).

The second subscript number, k, denotes the stage number *from* which the stream originates.

The third subscript number, j, denotes the product tree from which the stream originates.

By convention $j = k = 0$ will mean that the stream is a raw material stream originating from the environment.

In addition, when analysing the characteristics of a subsystem, the superscript (′) will indicate an output stream with respect to the analysed system. In this case the second and third subscript numbers are omitted because they provide redundant information.

Dimensions and units of stream flow rates

It is convenient to define dimensionless flow rates. For this purpose a reference stream is selected. It was found desirable to select the p_{nj} stream of each tree as the reference stream (e.g. the 'adopted product' stream originating from the consumer system or stage n of the tree).

The units in which a stream flow rate is measured are the dimensionless units obtained from the ratio of the stream flow rate in absolute units of quantity per unit time to the flow rate of the reference stream in the same units. Following this convention, by definition

$$P_{nj} = 1 \quad \text{for all } js.$$

It also follows that the dimensionless flow rate of the same stream may be different depending on the system which is considered in the analysis. This, however, presents no difficulty as will be shown.

For a certain type of dynamic analysis it is necessary to revert back to absolute units of flow rate. For example, when the effect of a change in the flow rate of the adopted product stream upon other variables of the system is studied, it is necessary to revert to absolute rate units.

A *component of a stream* is a definable quantity that characterizes a stream.

Composition of a stream is the set of all component quantities of a stream.

As in the case of flow rates which can also be considered to be a characterizing component of a stream, component quantities will also be expressed by dimensionless numbers. These are obtained by taking the ratio of the component quantity per unit stream flow rate to the component quantity in the reference stream per unit reference stream flow rate.

The stream components considered in a production system and in the present model are:

1. The number of man-hours of skill s invested in the stream at a certain point per unit flow rate. The symbol denoting this quantity in dimensionless units is x_{1s}. (Note that the subscript 1 corresponds to the class number of manpower subsystems in Figure 4.)
2. The number of equipment hours of type s invested in a stream at a certain point per unit flow rate. Symbol for dimensionless quantity: x_{2s}.
3. The number of energy units of type s per unit flow rate invested in a certain stream per unit flow rate. Symbol for dimensionless quantity: x_{3s}.
4. The number of 'service' units of type s invested in the stream up to a certain point per unit flow rate. Symbol for dimensionless quantity: x_{4s}.
5. Total cost invested in the stream at a given point per unit flow rate. Symbol for dimensionless quantity: x_{5s}.
6. Sales value of stream per unit quantity. Symbol: x_6.

The *system level* of an entity in a system is defined as the quantity of that entity in appropriate units which is in the system at a given time t. This quantity is symbolized:

1. By a Latin character (upper case) corresponding to the lower-case character characterizing a stream.
2. Three subscript numbers. The first i has the identical meaning as the corresponding one in a stream quantity. The second and third j and k refer to the stage and tree to which the level refers.

Levels are also measured in quantity units per unit reference stream flow rate. It should be noted that these units are dimensional and have the dimension of time. Component compositions will be symbolized by upper-case X with subscripts corresponding to the stream components.

Examples. R_{ijk} symbolizes the level quantity of the rejects of type i in stage k tree j measured in quantity units per unit flow rate of reference stream.

X_{1s} $\{R_{ijk}\}$ symbolizes the number of man-hours of type s in dimensionless units per unit R_{ikj} invested in R_{ikj} up to a certain time t. Under steady conditions X_{1s} is constant with respect to time.

Development of Mathematical Models of Production Subsystems

The following types of mathematical relationships are pertinent in defining the system operator:

1. *Balance equations.* These types of relationships are a detailed statement of the universal principle of conservation of entities. The logical process of deriving them is deductive in nature. Balance relationships follow the following format:

Rate of flow in = rate of flow out +
+ rate of accumulation − rate of generation.

2. *Dynamic or rate relationships.* As in the physical sciences, rate relationships are more difficult to formulate because they are inductive in nature. One has to rely on past experience in order to list the relevant variables, on correlative ingenuity in order to derive the mathematical functions, and on reliable measured data in order to compute and update the pertinent parameters of the functional expression. These relationships describe how the outputs vary with time when a certain time disturbance in any variable is introduced. The equations may be highly nonlinear. If they are linear they can be expressed in Laplace, Fourier, z, or other linear transform formats.

3. *Equilibrium or steady-state relationships.* These relationships describe the 'standard' steady-state condition. For instance, the following statement may be considered to be a stoichiometric equilibrium relationship:

> To drill a hole in a steel plate (a) $\frac{1}{50}$ man-hour, (b) $\frac{1}{20}$ machine time, (c) $\frac{1}{50}$ kilowatt-hour, (d) a total cost of 5 cents is required.

The parameters in the equilibrium relationships (to be referred to as equilibrium constants) are in most cases the integration constants of the dynamic equations. They are determined by various measurement techniques (e.g. work sampling, time study, etc.) and are a function of the nature of the operation, the nature of machines, the degree of knowhow (technical and work procedure), the organization structure, etc. If these variables vary at a high enough frequency, the equilibrium constants have to be reevaluated at a commensurate frequency.

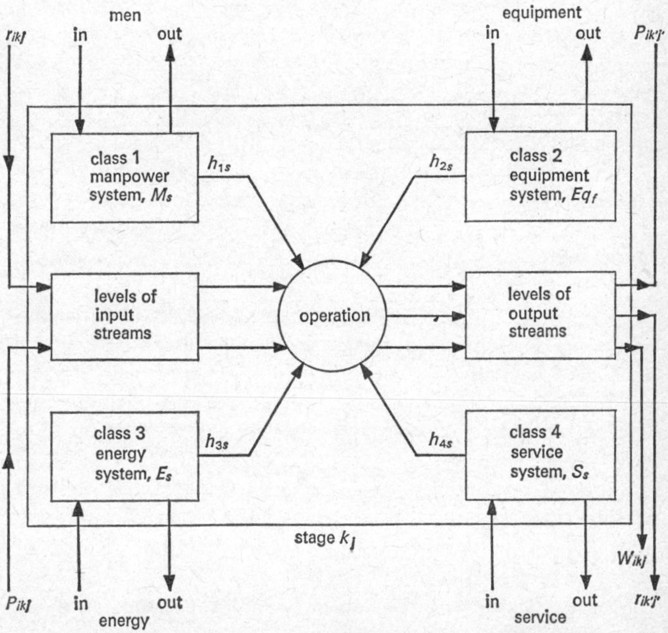

Figure 5 Stage k_i systems with subsystems, input–output streams

Equations for a Stage k_j Subsystem (Figure 5)

Balance equations

$$\sum_{j'}\sum_{k'}\sum_{i}(r_{ik'j'}+p_{ik'j'})=\sum_{i}(w'_i+z'_i+p'_i)+$$
$$+\sum_{i}\left(\frac{d(P_i+P'_i)}{dt}+\frac{d(R_i+R'_i)}{dt}+\frac{dW'_i}{dt}\right). \quad 5$$

Component balance equations

All the component balance equations have the same format. Below are the man-hour component equations:

$$\sum_{S}\sum_{j'}\sum_{k'}\sum_{i}(x_{1S}r + x_{1S}p)_{ik'j'} = \sum_{S}\sum_{i}(x_{1S}w'_i + r'_i x_{1S} + x_{1S}p'_i) +$$

$$+\sum_{S}\sum_{i}\left(\frac{dP'_i}{dt} + \frac{dR'_i}{dt} + \frac{dW'_i}{dt}\right)X_{1S}$$

$$= -\sum_{S}\sum h_{1S}\left(r'_i + p'_i + \frac{dP_i}{dt} + \frac{dR_i}{dt}\right) \qquad 6$$

and similarly for other components.

The terms involving differentials on the right-hand side of equations **5** and **6** disappear for steady-state conditions.

In equation **6**, h_{1S} is the number of dimensionless man-hours added within the system per unit $r'_i + p'_i + dP_i/dt + dR_i/dt$. Similarly defined symbols are h_{2S}, h_{3S}, h_{4S}.

To illustrate the usefulness of the proposed framework consider a production system for two consumer products. Figure 6 shows schematically the two trees defused by the two products 1 and 2. Product tree 1 is assumed to have six stages including the distribution and consumer systems. Product tree 2 is assumed to have four stages.

The area marked $2'_1$ in Figure 6 is not considered to be a separate stage, because the only input and output streams to and from it originate and go to adjacent stages with the exception of the waste-product stream.

Now we could write the balance equation for stage 4, as an example of applying equation **5**.

$$r_{151} + \sum_{i}(p_{i0} + p_{i31} + p_{i12}) = p_{141} + \sum_{i}(r'_{i41} + r_{i41} + w_{i41}) +$$

$$+\sum_{i}\left(\frac{d(P_i + P'_i)}{dt} + \frac{d(R_i + R'_i)}{dt} + \frac{dW'_i}{dt}\right). \qquad 7$$

Note also that

$$p_{141} - r_{151} = p_{161} \quad \text{(adopted product).}$$

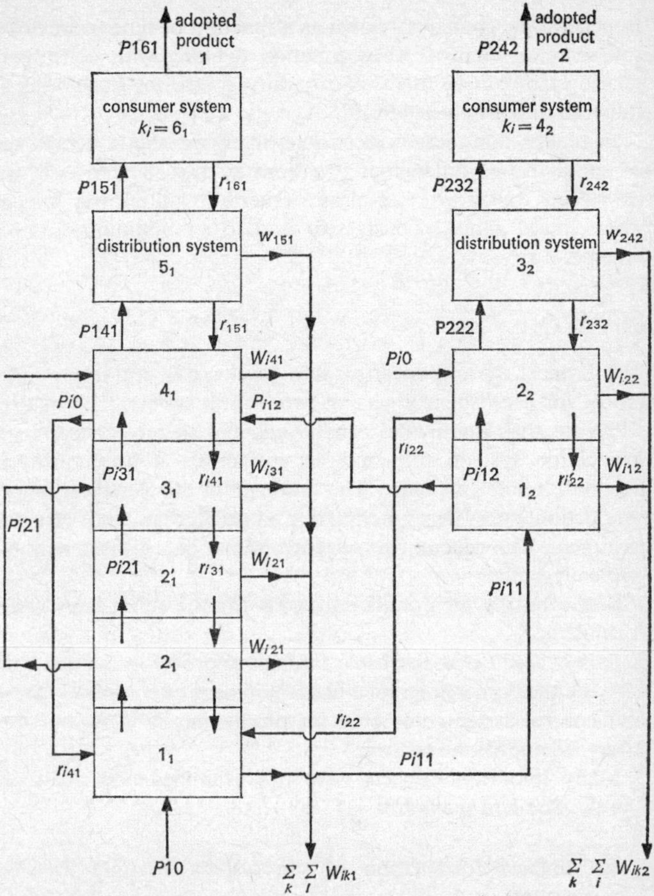

Figure 6 Production system for two end-products

Thus

$$p_{161} = \sum_i (p_{io} + p_{i31} + p_{i21}) - \sum_i (r_{i41} + r_{i41} + w_{i41}) - \\ - \sum_i \left(\frac{d(p_i + p_i')}{dt} + \frac{d(R_i + R_i')}{dt} + \frac{dW_i'}{dt} \right). \quad 8$$

In general, p_{161} must be known as a function of time from close consumption control. Thus equation **8** along with constraint relationships can be used for inventory control of stage 4, as a function of consumer demand.

In reality, considering the whole plant, we obtain a number of simultaneous differential equations of type **8** and a set of constraint equations and profit criteria relationships, which when solved simultaneously will lead to an optimum control scheme.

Conclusions

In this paper the authors attempt to develop a logical framework model for a completely general production system. The reader will note that this model can be adapted to any conceivable production system, for example, a factory, a warehouse, a hospital, a hotel, an office. The rationale of the classification of production subsystems permits a most flexible and general treatment. The general problem of optimal control was mathematically stated.

Some, but by far not all, equations for the subsystems were formulated.

It is believed that the basic rational approach presented can serve as the cornerstone for the development of completely general control systems not only for production systems, but for larger organization systems.

Many important aspects were not examined and analysed closely. Some of them are:

1. Complete formulation of the system dynamic transfer operators.
2. Analysis of the control system in relation to the production system.
3. Development and formulation of the profit function.
4. Analysis of errors in the measurement and other subsystems.
5. The system design problem.
6. The dynamic effect of psychosociological factors.
7. The mathematical problems of solving the highly complex relationships.

These are only a few of the many questions that deserve detailed and careful analysis which may provide the challenge to many years of creative effort.

Glossary of Symbols

Symbol	*Description*	*Equation*
$u_i(t)$	Input variables	1
$v_i(t)$	Output variables	1
M	System operator	1
Q	Objective function	2
B	Profit function	2
C_j	Constraint function of the jth kind	3
p	Product stream	
r	Reject stream	
w	Waste stream	
X_{1S}	Man-hours of type S per unit flow rate	
X_{2S}	Equipment hours of type S per unit flow rate	
X_{3S}	Number of energy units of type S per unit flow rate	
X_{4S}	Number of 'service' units of type S per unit flow rate	
P	Product level	
R	Reject level	
W	Waste level	

	Subscripts	*Equations*
i	Denotes classification of stream	5, 6
j	Denotes product tree number	
k	Denotes stage number	
i'	Denotes $i' \neq i$ (different from i)	
j'	Denotes $j' \neq j$ (different from j)	
k'	Denotes $k' \neq k$ (different from k)	

	Superscripts
I	Output stream

References

BUFFA, E. S. (1961), *Modern Production Management*, Wiley.
FORRESTER, J. (1961), *Industrial Dynamics*, Wiley.

20 S. E. Elmaghraby

Computers in Production Systems

Excerpt from S. E. Elmaghraby, *The Design of Production Systems*, Reinhold, 1966, pp. 415–45.

Simulation

Introduction

In the past few years, simulation has come to occupy a prominent position as a tool of design and problem solving.

To *simulate*, according to *Webster's Collegiate Dictionary*, is 'to feign, to attain the essence of without the reality'. In substance, therefore, every representation is a form of simulation and must involve some reduction or change from reality. The degree of difference between the simulation and the real system (real in the sense of capability of construction, not present existence) depends, of necessity, on the objective of simulation. In particular, the differences lie in those aspects of the real system which cause it to be real and which are irrelevant to the study.

An example should clarify this rather lengthy definition. To the supplier of army uniforms, the sizes of uniforms that he makes should conform to some statistical distribution, which is the same distribution of the sizes of men recruited in the army. Now, if we need to conduct some simulation studies on these uniforms, say an inventory study, the system that we would be investigating would be composed of, among other things, random numbers that conform to the specific statistical distribution. This is the *essence* of the real-life situation of interest to us, and it is naturally the feature that we must retain in the simulation model. The other features that make real life situations 'real' are irrelevant to our study. Thus, the names of the army personnel requisitioning the uniforms from the storeroom, their religion or political affiliation, their disposition towards army life, or their ideas on current social problems are completely irrelevant facts. Some of these facts may be relevant to other studies.

Simulation usually involves a great number of calculations, hence the common phenomenon of using computers in simulation. However, it is important to bear in mind that large scale digital (or analog) computers are not *necessary* to conduct simulation studies – a great deal can be done and has been done with pencil and paper and with the possible aid of a desk calculator. [...]

Simulation may or may not involve *Monte Carlo* experimentation. In other words, the functional relationships among the various components of the model may or may not involve a random element which is subject to some known probability law. For example, one can simulate the flow of jobs in a manufacturing shop assuming a completely deterministic system, with the object being the experimental determination of the distribution of completion time as a function of the job sequence. Whenever Monte Carlo simulation is adopted, sampling from the populations of the random variables is necessarily used. Population sampling also extends to sampling attributes and events whenever complete enumeration is not feasible or not possible, such as in the simulation of the behavior of a job shop mentioned above.

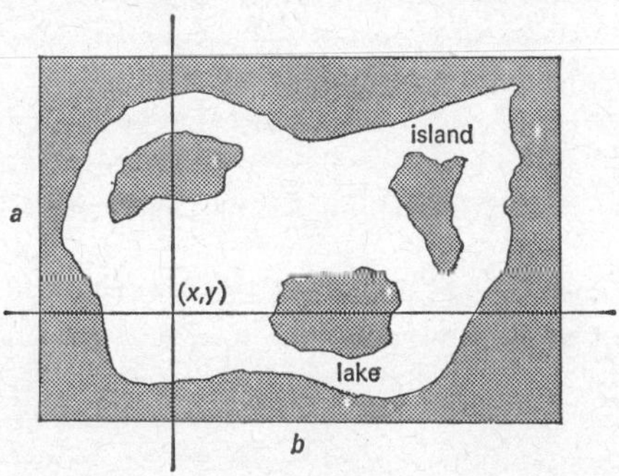

Figure 1 The problem of finding the area of an odd-shaped lake

But population sampling may also be used as a problem-solving technique of its own right. Consider, for example, the following problem. Suppose it is desired to determine the surface area of a lake which is of irregular shape and contains several islands, as shown in Figure 1. Instead of engaging in a frontal attack on the problem, let us reconstruct it in the following manner. Enclose the lake in a rectangle of sides a and b. Now, any point (x, y) chosen at random is either on land or in the water. The proportion of points falling in the water approaches the relative water area to the area of the rectangle as the number of such sample points grows large.

In summary, then, simulation is a problem-solving technique, the success of which relies on the representation of the essence of the real-life problem, and which may or may not involve Monte Carlo experimentation. In the following sections we will take a closer look at the technique itself.

Why simulate?

All of the above discussion leads to the conclusion that simulation is a *technique*. It is not a theory; rather, it is a problem-solving methodology. In fact, it is one of many problem-solving techniques available to the systems engineer. It need not be the best technique for *all* problems. In fact, oftentimes it is a very poor technique indeed.

This naturally brings up the question: when is simulation useful as a problem-solving technique? We mention the most prominent among the reasons for simulation, which may also serve as a list of the *uses of simulation*.

1. There is an absence of any compressed and complete mathematical formulation of the problem, or great difficulty is encountered in *solving* such formulation if one exists. The complexity of the system may be the reason for either of these eventualities. For example, to date it has not been possible to reduce the operation of a large business firm to a neat set of mathematical equations.

2. Scientific investigation proceeds from observation to hypothesis to experimentation and finally, to validation. Simulation may be the *only possible way* of experimentation because of the impossibility of conducting the experiments and observing the phenomena in the actual environment. This is true, for example, of studies of the thrust of rocket motors for use in interplanetary space. Alternatively, simulation may be the *cheapest* way of conducting such experiments when compared with experimentation with the real system, a pilot plant, or even

in the laboratory. This is the case in the determination of a new warehouse location, for example.

Simulation can be considered the ultimate in controlled experimentation since even random or chance variability can be replicated with the greatest of ease. The statement 'with all other factors remaining constant' can be made literally true. This absolute control of the experiment may be a very desired feature, but it does raise some problems of comparative validity which we shall discuss below under problems of simulation.

3. A simulation is a very powerful *educational* device. The very act of constructing a simulation is instructive to all participants, since complete familiarity with the simulated system is a prerequisite. It is a very good *training* device, especially if combined with some physical representation of the system. To be able to 'see' and 'play with' the system, that is to say, with its simulation, is of tremendous value in (a) gaining insight into the relative importance of the various factors, (b) perceiving the system as a whole with all its interacting elements, (c) selling new ideas and gaining enthusiasm for proposed changes.

4. Simulation may actually assist the process of innovation and discovery by imparting such 'feel' for the problem as mentioned in (3). The output of simulation is data, and data, whether real or artificial, may suggest new questions or avenues of research not envisaged at the start of simulation. More than one instance is on the record of an investigation which started by asking certain questions only to be diverted more than once to more interesting, and oftentimes more fundamental, questions.

The power of simulation in this respect is apparent in the study of new policies or in acquiring new equipment. *Before* the new policy is instigated or the equipment installed, the simulation can answer many questions, point out unforeseen problems, and indicate the relative merits and demerits of alternatives. This may avoid a great deal of expense and disillusionment later on.

5. Besides the complete control of the experimental environment mentioned above (2), simulation affords complete control over *time*. This immediately yields two results: (a) a sort of 'slow motion' study of fast-acting events can be made (as well as a 'speed up' of slow-moving events), and (b) one can handle

several projects moving at different rates, such as in attack-defense maneuvers.

6. Finally, simulation can be used as a check on analytical solutions. To quote from Morgenthaler (1961, p. 374):

> In a recent missile study, the author arrived at a closed-form expression for the approximate expected damage to a circular target area due to an impacting circular cluster of warheads whose mass center was normally distributed about the target center. There were no tabled values against which to check the accuracy of the approximation. It seemed desirable to 'check' the formula against some known cases before employing it widely in the Operations Research study. Using cut-out paper damage circles, a circular target, and a table of random normal deviates, it was possible to obtain 'experimental' data which checked out the analytic result.

Having discussed at some length the reasons why simulation is 'good', it would seem only fair to give the reader the other side of the coin and draw up a list of why simulation is 'bad'. A brief list of the limitations of simulation is given on page 354. In order to grasp the full significance of these limitations, we will take a slight detour in the next two sections to discuss the construction of a simulation and the importance of simulation languages in facilitating such construction, respectively. Then we shall return to the subject of problems in simulation and the limitations of the technique.

On the construction of a simulation

How does one obtain a simulation of a complex system? There is no uniform step by step method recommended for the construction of a simulation, but since a simulation is a *model* of the system, it must follow that the approach to the construction of a simulation is identical to the construction of a model. In particular, each component of the system is represented by a mathematical model, and the interactions among the components are specified. A computer program is then written to perform in concordance with the mathematical model. When a simulation is 'run', the program proceeds step by step through short intervals of time, pausing at the end of each interval to perform the indicated functional response of the components and to evaluate the status of the system. In the process of so doing, various statistics and measures of performance are calculated. Thus, it is possible to obtain the history of the performance of the system under any given conditions. By repeating the experiment under different conditions – different either in a deterministic or probabilistic fashion – it is possible to compare results and decide upon a 'better' set of decision rules.

The following are a few remarks on the methodology of constructing a simulation.

First, it is important to *possess detailed acquaintance* with each component of the system and its manner of interaction with other system components. If the smallest elements into which we can divide a system are themselves of unknown behavior, either deterministically or probabilistically, simulation is not feasible.

Second, detailed and precise *definition of the nature and scope of simulation* must be made as early as possible. There are strategic as well as tactical considerations in this respect: (a) the objective of the simulation, (b) the sources of input data, their form and content, (c) the financial, computer, and manpower support available or necessary, (d) the duration and extent of the simulation, and (e) the experimental design that will yield the desired information, and the manner in which each 'test run' is to be executed.

Third, it must be evident that one advantage of simulation is that it *requires no simplifying assumptions*; the model can be made as complicated as desired. This advantage can be easily turned into a disadvantage by the 'dressing up' of the simulator to give it the 'appearance of reality'. This may prove, in the end, to be the death trap of the simulation and is too often the case to be dismissed lightly.

The reason for such an unpleasant ending is simply the following. A simulation is a model that is 'run' rather than 'solved' in order to obtain results. Consequently, by its very nature, simulation is incapable of generating its own solutions, and is, therefore, confined to the analysis of the behavior of the system under any specified conditions. In short, simulation is an 'if-then' tool of analysis. It immediately follows that simulation, when not used as a mere descriptive tool, is best suited for *comparing alternatives.* If the number of factors affecting the result under any given alternative is large, a valid comparison among alternatives can be made only after each alternative has been tested under *all possible combinations* of the interacting factors. This can be a very large number indeed, and oftentimes it is of such astronomical magnitude that complete factorial experimentation is ruled out from the start. Of course, design models other than complete factorials exist and do permit valid comparison among alternatives. However, the number of necessary experiments under such designs may still be inordinately large if the number of variables becomes excessive. In stochastic processes, the problem is further complicated by the necessity of taking representative random samples of the different measurements.

The price of encompassing *all* factors affecting the real life system, and therefore being 'realistic', is an unwieldy simulator and a mass of data that does not mean a thing. On the other hand, a simulation which ignores many factors may lead to erroneous conclusions. The art of simulation lies in balancing these two extremes and in coming up with meaningful results.

Apart from the question of factorial interaction, which is a question of size of experimentation, there is the question of *abstraction* and *reduction*; is the simulation going to be the same *size* as the real system? This is a question of *content* of the simulation proper. For example, suppose it is desired to simulate a shop containing 400 machine tools of different capabilities. Should the simulation contain exactly 400 machine tools, or would a smaller number suffice? What is the 'essence' of the problem here? In particular, would fifty machine tools do? How about *three* machine tools?

Almost every analyst who is concerned with simulation will, sooner or later, be faced with the question of the extent and validity of abstraction and reduction. It is not possible to give a hard and fast rule in this respect, since the question of validity of representation depends on the real system being represented. Limiting cases may or may not contain the essence of the complete system, and if they do not, it would be a mistake to generalize from the limiting case to the parent system.

For example, a point is a circle of radius zero; it is a limiting case. And yet, no amount of study of the properties of the point will reveal anything concerning the properties of a circle of radius greater than zero. Or, consider the following example from economic theory. The so-called 'Robinson Crusoe economics' is the economics of one man, assumed to be a 'rational economic man', against 'nature'. The entrepreneur, which is the usual decent name given to such a man, is out to maximize *his* utility. The 'Robinson Crusoe economics', which dominated economic thought for over two centuries, is the limiting case of real life economics and is a valid model of monopolistic economy. But it is *not* a valid representation in the case of duopoly (or two giant firms controlling the market). There is a fundamental element which is present in duopolistic markets which is completely absent in the Robinson Crusoe theory. That element is *competition*, and it took the genius of von Neumann to develop the mathematical theory of two-person games to represent such a system.

But even the two-person game theory falls short of representing the market under oligopoly (or a few firms controlling the market). There is an added element which is present in oligopoly but is completely absent in both the Robinson Crusoe theory or the two-person game theory. That element is *collusion*. Again, it took the genius of von Neumann to develop the n-person game theory. It is instructive to note that the three-person game theory contains *all* the elements of the n-person situation, and it is amusing to remark that as n grows large, the theory reverts to the Robinson Crusoe economic theory!

Thus, we see that no amount of study on the one-machine shop will reveal anything concerning the many-machine shop, except perhaps the negative result that one cannot generalize from the one-machine study to the many-machine case.

In general, as was stated before, the level of abstraction and the degree of reduction are functions of the real system under study. Achieving the *essence* of the real problem must not be confused with achieving the *like* of the real problem.

One final remark concerning the construction of a simulation: in many instances it may not be feasible to carry out a frontal attack on a problem, or it may be preferable (from time, effort, or other points of view) not to carry out such an attack. Rather, it may be possible to solve easily *another problem* whose solution gives the result sought in the first place. This is a standard 'trick', and a rather clever one, in mathematical and engineering analysis in general. [. . .]

We encountered such a 'trick' in the previous section when we wanted to calculate the surface area of an odd-shaped lake. There, we solved a probabilistic problem (which was, in effect, the determination of the binomial parameter p denoting the probability of the sample point falling in the water area) whose solution also yielded the surface area of the lake.

In many simulation experiments, the well-known ergodic principle is invoked: instead of determining the desired statistical characteristics from the study of a very large number of events occurring at the *same time*, we observe the behavior of a few events *over a very long period* of time. Sometimes this is much easier to do, and the results are equivalent.

Simulation languages

Introduction. Readers of cartoons and science fiction are familiar with computers that allow their employers to write their problem in vague terms on a slip of paper and drop it into a slot. The computer then proceeds to instruct everybody around on what *they* should do. Unfortunately, or perhaps fortunately, this blissful state of affairs has not been reached yet, and it is most likely, at least for some time to come, that man will be instructing the computer on what *it* should do. In fact, these instructions must be sufficiently complete and precise so that they leave absolutely nothing to the 'discretion' of the computer.

The communication between man and machine, just as the communication between one man and another, is effected through the medium of a *language*. It is the purpose of this section to discuss computer languages, particularly as they relate to the technique of simulation.

To this end, we need to review a few concepts widely used in the field of computer data processing. *Programming* encompasses three main activities. They are: block diagramming, coding, and 'debugging'. A *block diagram* is a model of the behavior of the system and expresses, in graphical form, the logic of the system. A block diagram thus serves the dual purpose of representing the system and specifying the desired logic of operation according to which the computer should conform. A block diagram is usually developed in several stages, beginning with a gross over-all view of the behavior of the system and ending with a detailed specification of the logic of operation.

The second function, *coding*, entails the detailed specification of the location and form of input and output in memory, the allocation of memory among the various portions of the program (input-output, reference tables, working storage, and the program itself), and the detailed translation of the block diagram logic into machine-understandable language.

Finally, '*debugging*' is the common term given to discovering and correcting the errors, logical and otherwise, in the previous two functions.

Our main concern here is with the second function, coding, or, in general, the problem of preparing a machine-understandable program.

It became evident in the very early stages of computer development that the chore of coding in basic machine language weighs heavily on those who wish to utilize computers, and that the work involved in coding acts as a serious inhibitor to the wide usage of computers in the various fields of science, engineering, and business. In particular, the user usually thinks in a language which is very different from the basic machine language. For example, a mathematician thinks in terms of mathematical symbols and notation, while a businessman thinks in terms of common English language. The basic language (where a 1 may stand for 'addition' and an asterisk for a drum location) is really a nonsense symbol language from the user's point of view. Effort for the past two decades has been directed towards the elimination of this barrier by devising 'higher languages' which more and more approximate the language in which the user usually thinks. The fruits of such efforts show in *assembly* programs, *interpretive* routines, and *compilers*. All three are symbolic programming systems which can be classified as attempts towards *automatic* programming.

An *Assembly Program* (or Symbolic Program) permits the user to write his commands in pseudo-instructions, where a pseudo-instruction is usually mnemonic for the function to be performed. For example, in the I.B.M. 1401 computer system,

A stands for 'add', and ZA stands for 'zero and add', and so forth. The assembly program then proceeds to translate from these pseudo-instructions to machine language instructions. It also assigns actual machine memory addresses for 'names' of data or instructions and translates from decimal to binary notation. Coding in pseudo-instructions for use on assembly programs represents the first level of departure from direct machine-language coding.

The second level of development is, naturally enough, to make one pseudo-instruction represent several machine language instructions. For example, instead of programming the various operations performed in taking the square root of a number, which may result in ten separate instructions, it is sufficient to indicate in one instruction that one wishes to evaluate the square root. The assembly program would then analyse the pseudo-instruction, construct the necessary machine instructions by reference to standard subroutines and end up with the desired machine language instructions. In this fashion, the programmer is asked to write in *macro-instructions*; the assembly program translates into *micro-instructions*.

There are two modes of operation at this level of abstraction. In the first mode, *interpretive routines* act upon the macro-instructions as they appear in the program. A supervisory program calls in the proper subroutines *each time* the pseudo-instruction is executed. This is wasteful in operating time and in machine memory capacity because all the subroutines as well as the supervisory program must be in the computer all the time. However, the necessary coding instructions are extremely easy to learn, and computer operation is very elementary indeed. This mode of operation was utilized extensively in simple engineering and scientific work, where individuals were permitted direct access to the computer to evaluate simple problems, but is now almost obsolete.

The second mode of operation is through a *compiler*. The action of a compiler is similar to interpretive routines except that the pseudo-instructions are 'compiled' into a machine language program once and for all in a separate machine run. In performing the act of translation, the compiler also allocates machine memory to the various components of the program.

Compilers have distinct advantages over other modes of operation. Among these are the following: (1) Since the whole computer memory and logic is available to the compiler during the translation run, a more sophisticated language may be used which more closely approximates the language in which the user thinks. (2) Compilers have built-in error-detection subroutines which, together with the resemblance of the programming language to the user's own thinking language, render correction and 'debugging' a much easier task indeed.[1] (3) Compilers provide a common language to the users of various computing equipment.

The major disadvantages of compilers are the following: (1) The possible inefficiency in their inability to capitalize on individual machine characteristics. This manifests itself in longer processing time compared with a well-written symbolic assembly program. (2) Loss of access to certain logic features of the equipment; e.g. examination of the bit structure of a character. (3) Less efficient use of memory because of various reasons such as use of memory for program linkages, absence of tight packing of words, etc.

The most widely used compilers are FORTRAN (Formula Translator), basically a language for scientific and engineering systems, and COBOL (Common Business Oriented Language), which is, as the name indicates, primarily for business activities.

It is not difficult to project into the future of computer languages. The trend is toward higher and higher levels of abstraction, so that the movement is from macro- to macro-macro-instructions. This is especially successful if the language is restricted to a specified field, such as structural engineering design or manufacturing processes. In the following, we will study the development of such specialized languages in a very particular field, the field of system simulation.

Advantages of simulation languages. The problem of preparing a simulation falls into two tasks. First to construct a model of the system, and second to program such a model for computer input and manipulation. Simulation languages are

1. Advantages 1 and 2 of compilers are also shared by Assembly or Symbolic programs.

intended to facilitate the second task by reducing by a considerable measure the amount of effort expended between the construction of the model and the availability of a computer program for simulation. This is accomplished through a language which is a *description of the model* independent of the particular computer being used. The analyst is required to be familiar only with the system and the descriptive language of the model, leaving the question of detailed conversion into a computer program to a compiler program which is responsible, later on, for such conversion into detailed computer instructions. Needless to say, such a simulation language relieves the analyst of the major burden of familiarity with the simulated system *and* with the computer which is used for running the simulation.

The major difference between simulation languages and the general purpose scientific or business languages mentioned above lies in the specialization of the former, which permits them to be more powerful though of a much more limited 'vocabulary'. Thus, power and subsequent saving in programming time and effort are bought at the expense of flexibility and adaptability to a variety of problems.

Currently, there are two simulation languages which are predominantly in use: G.P.S.S. (General Purpose Systems Simulation) and SIMSCRIPT. The two languages are very different in structure, and the reader whose duties necessitate the simulation of systems is urged to familiarize himself with the power and limitations of *both* languages.

Other simulation languages have been developed,[2] and undoubtedly many more will be forthcoming. It is highly improbable, however, that there will be any radical departure from the concepts of these two languages unless a new generation of computers is developed with radically different logical structures.

We illustrate the concept of simulation languages with an example of a simple manufacturing process using G.P.S.S.

G.P.S.S. (General Purpose Systems Simulation). This language was developed by Geoffrey Gordon (1962) of I.B.M., on whose various writings we shall rely in this brief resumé. The utility of

2. For example Mathematics and Applications Department, I.B.M. (1960); Kavanaugh (1961, pp. 15–22).

G.P.S.S. has been largely in simulating traffic systems, such as communications, manufacturing, and air terminal control systems. In such systems the basic unit of interest, called *transaction*, is followed through the system and its movement studied with respect to path, congestion and occupancy problems. The physical components which operate on the transactions are called *items of equipment*. An item of equipment is either of unit capacity, such as a machine, which can accommodate only one transaction at any time, and will be called a *facility*. Or, the item of equipment is of multiple capacity, such as a multi-channel switching system, and will be called a *store* whether or not it is in fact used for storing.

Similar to any other language, G.P.S.S. has a fixed set of symbols together with a 'dictionary' giving the meaning of each symbol, and a 'grammar' specifying the usage of each symbol. The symbols are geometric shapes: circles, rectangles, trapezoids, etc., and there are thirty-five such symbols. Five examples of such symbols are shown in Figure 2.

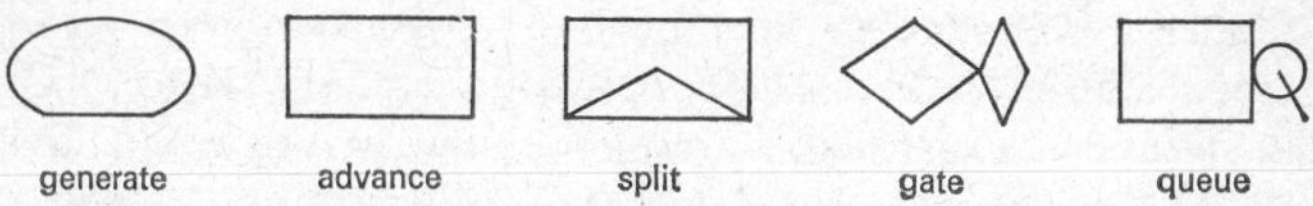

Figure 2 Examples of G.P.S.S. symbols

The simulation program specified by the analyst is a *block diagram* constructed in a specific fashion. Each block of the diagram represents a step in the action of the system, and lines joining the blocks indicate the sequence of events. To represent the alternative courses of action that may be followed by the system, more than one line may leave a block. Correspondingly, one block may be entered by way of more than one line to indicate that the step represented by that block occurs in more than one sequence of events.

The block diagram is distinguished by two aspects. First, each block is given a number which serves as its 'name'. Second, each block representing an action which consumes time is given a *block time* which may be a fixed integer, a random variable, or a function of either.

All block types except terminal points of the system may have two exits. The exits are distinguished by referring to them as exits *A* and *B* and they are defined by giving the number of the block to which the exit leads; these latter are called *next blocks A* and *B*. The routing of the transaction may be specified in either conditional or unconditional forms. For example, a *conditional* mode of selection would be to attempt exit *A* first, and if the transaction is refused entry to next block *A*, it will attempt to exit via next block *B*. If neither exit is free, the transaction will be delayed and will leave by the exit which becomes free first. If both become available simultaneously, exit *A* is chosen. This pattern may simulate preference in actual behavior between the two exits. On the other hand, an *unconditional* selection may behave in the following fashion. A random selection with equal probability is made among all next blocks whose numbers fall between next block *A* and next block *B*, inclusive. If exit is not permitted because of occupancy of that particular block, the transaction waits until the block is free. For example, if next block *A* is number 20 and next block *B* is number 29, the simulator will choose with probability 1/10 among the blocks number 20, 21, 22, . . . , 29. The block selected is the one to which the transaction is routed. Such a model may be a representation of the random selection by arriving customers from among a given set of service facilities.

The movement of a transaction from block to block constitutes an *event*, which occurs at some point of time. The simulation program maintains a record of the times at which these events are due to occur, and proceeds by executing the events in their correct time sequence. In case any transactions cannot move because of some obstruction in their paths, the program moves them as soon as these obstructing conditions are removed (which movements themselves would constitute events.) *Time* is measured on a *master clock* which is internal to the simulation. The unit of the *clock time* is specified by the programmer and constitutes the basic unit in which all system time measurements are taken.

The program proceeds from the time of occurrence of one event (or events, if all occur at the same clock time) to the time of occurrence of the next event(s). The block time is thus advanced by the elapsed time between the two occurrences.

In a simulation, *structure* is obtained by specifying the logical relationships among the variables. G.P.S.S. provides the symbolism required for the specification of such decisions. For example, HOLD, SEIZE, RELEASE, INTERRUPT, RETURN specify a variety of such decisions either on the item of equipment or on the transaction.

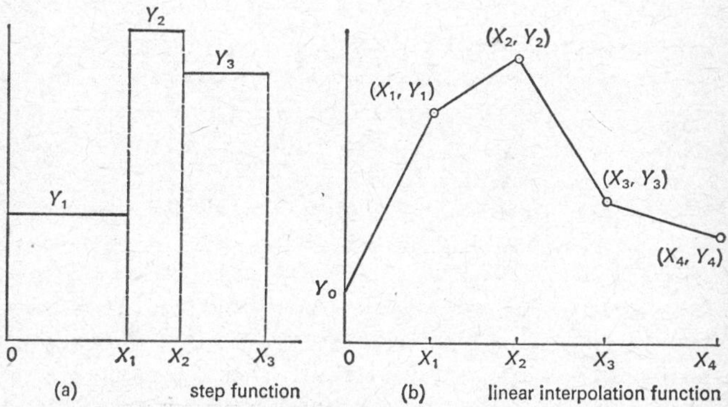

Figure 3 The two types of functions

Moreover, simulation is conducted to accumulate some statistics on the various conditions of the system. For example, we may be interested in the proportion of time a facility is utilized, or perhaps in the degree of its utilization, or in the amount of congestion at a given facility, etc. Again, a number of blocks give such information directly: QUEUE, MARK, TABULATE, etc.

Finally, the G.P.S.S. language provides for the use of *functions* in the various aspects of simulation. For example, it was mentioned before that the *block time* may be a function of a constant or a function of a random variable. Or, the selection of the next block may be a function of some condition of the system, and so on. In G.P.S.S., a function is a numerical value computed from a table of numbers defined by the user, and there are two types of functions: step functions and linear-interpolation functions, Figure 3. The first type, as in (a), is given by

$$(x) = y_k \quad \text{for} \quad x_{k-1} < x \leqslant x_k \quad (k = 0, 1, \ldots, n).$$

The second type, as in (b), is given by

$$(x) = y_{k-1} + \frac{y_k - y_{k-1}}{x_k - x_{k-1}}(x - x_{k-1})$$

$$\text{for} \quad x_{k-1} < x \leqslant x_k \quad (k = 0, 1, \ldots, n).$$

The function is indicated to the program by designating its kind and specifying the co-ordinate points (x_k, y_k).

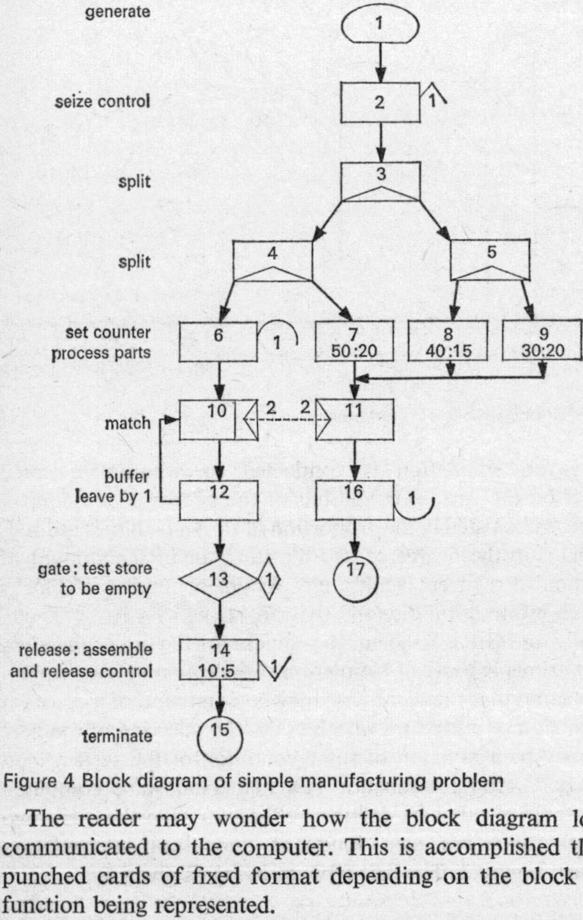

Figure 4 Block diagram of simple manufacturing problem

The reader may wonder how the block diagram logic is communicated to the computer. This is accomplished through punched cards of fixed format depending on the block or the function being represented.

To summarize, G.P.S.S. views the real system as composed of

facilities and *stores* through which *transactions* pass following some logical decisions. The behavior of these transactions as they filter through the various components of the system is the subject of the simulation. Representation of the system and the interaction among its components are accomplished by *block diagramming* the system according to a well-specified language.

Instead of proceeding with the detailed description of G.P.S.S. language, giving its dictionary and grammar, we will illustrate its usage by a simple manufacturing problem (Gordon, 1962, p. 31). This should give the reader an idea about the language and its utility. But, like any other language, mastery can be gained only through practice. The reader who wishes to pursue the subject further may refer to the complete manual on G.P.S.S. written by G. Gordon and listed at the end of the article. Other examples of the use of G.P.S.S. can be found in the literature listed at the end of the article.

The example shown in Figure 4 represents a simple manufacturing process in which one job at a time is processed. The job requires that three parts be made and these are made concurrently. When all parts are ready, they are assembled and the next job is started.

In order to understand the complete block diagram, we first define the blocks used in it and indicate their usage.

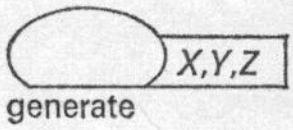

The GENERATE block creates transactions. The rate of flow is specified in terms of an interarrival time between successive transactions. The X value specifies the entry time of the first transaction; the Y value specifies the time between the creation of transactions; and the Z value specifies the priority of the transaction (any number between 0 and 7, inclusive).

The execution of the GENERATE command is in the following fashion. When the program encounters the GENERATE block it immediately creates a transaction. The transaction is scheduled to enter the block after an interval that is computed from the X and Y values. For example, if $X = 1000$, and $Y = 10$, the first transaction is created and is scheduled to enter the GENERATE block at clock time 1000. At that time, the transaction will be placed at the GENERATE block. The next block is then selected for the transaction, according to the selection mode at the GENERATE block. If successful, it enters, and the GENERATE block becomes empty. If unsuccessful, it *stays* in the GENERATE block. If this latter case happens, at time $X + Y = 1010$ a new transaction will *not* be created. At time 1020, a new transaction will be created only if the delayed transaction in the GENERATE block has succeeded in exiting in the interval 1010–1020, inclusive. Thus, delays at the exits of a GENERATE block reduce the rate at which transactions enter the block diagram. The queue in the GENERATE block will never exceed 1, even though the specified interarrival time (10 units in our example) may be less than the service time on the exit facility.

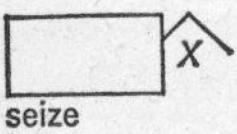

The SEIZE block records the usage of a facility, whose block number is given by X, by a transaction that enters the block. The facility will not be returned to available status except when the transaction enters, later on, a RELEASE block. When a facility is seized, any other transaction will be refused entry to the SEIZE block. The fate of that rejected transaction is not specified at the SEIZE block, however, but at the block from which the transaction is attempting to exit (a GENERATE block in our example). An action time at the SEIZE block is computed in the usual fashion. If no action time is given, it is interpreted as being zero. The advantage of this SEIZE–RELEASE setup is the ability to evaluate the time a transaction consumed seizing a facility.

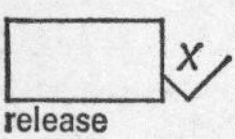

The RELEASE block performs the operation required to place the facility indicated by the value of X, in the 'available' status. It counteracts the action of the SEIZE block. In fact, a transaction may not release a facility unless the facility has already been seized by the *same transaction.* A host of system statistics is updated at the RELEASE block, such as duration of seizure, cumulative seizure of facility, average occupancy, etc.

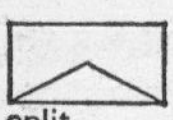

The SPLIT block creates a duplicate of each transaction that enters the block. The SPLIT block never refuses entry to a transaction. An action time is computed for the transaction. The new transaction receives the same parameter values, priority and transit time as the original transaction. In addition, both transactions receive an identifying code which will permit the program to search for other members of the same set of transactions. The next block is selected for each of the transactions in the following manner: the original transaction is sent to next block *A* of the split block, and the duplicate is sent to next block *B*.

advance

The ADVANCE block performs two actions: (1) computation of the number of clock units which a transaction is to spend at the block, and (2) selection of a block to which that transaction will be sent after the computed interval has elapsed. The principal use of the ADVANCE block is to represent an event in a system that requires a period of time or choice of paths but does not alter the status of an item of equipment. The duration of processing is usually given by the mean and the spread, which are written inside the block thus, *m*:*s*. As indicated before, there are alternate methods for specifying the duration of processing, but we need not go into them here.

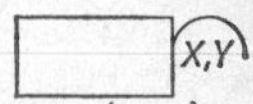

enter (store)

The ENTER block is one in a family of three, which are: STORE, ENTER, and LEAVE. The STORE block allows a transaction to occupy space in the store specified by the value of *X* for as long as the transaction is in the block. The ENTER block allows a transaction to take up space in the store but the space is not given back until a transaction (not necessarily the same one) enters a LEAVE block.

The *Y* field of the ENTER block (as well as of a STORE block) specifies the number of spaces needed for the transaction in the storage. Needless to say, this number need not be unity. The program compares the requested number of units of space with the unused space. If the number of units requested is not greater than the number currently available, the following steps are executed. Otherwise, the transaction will either be delayed or diverted, according to the selection mode and exits at the block from which the transaction is attempting to exit. Incidentally, when the space requested is a parameter, and therefore each transaction may call for a different number of units of storage, it is possible for a later transaction requesting fewer units to pass a delayed transaction which requested more units than were currently available in storage. When the transaction is admitted in the store, the program updates the storage content, computes time of occupancy, and selects a next block in the usual fashion. If the transaction cannot leave the selected exit or exits, no storage units are released.

leave

The LEAVE block is used to make available some previously occupied space in an ENTER block. The action of a LEAVE block is the reverse of the ENTER block; *X* specifies the store, and *Y* specifies the units of space relinquished. Certain statistics of interest to the simulation are computed such as: length of time for which the storage has remained at its current contents, a count of the total number of units that have passed through the storage, etc.

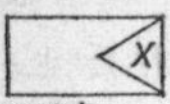

match

The MATCH block serves to synchronize the progress of two transactions of an assembly set. Instead of joining the pair of transactions, the MATCH block permits them both to continue advancing independently through the block diagrams, but the realization of a match permits other events to occur. The operation of a MATCH block is as follows: The transaction that is attempting to enter a MATCH block is removed from its previous block and a next block is selected for the transaction. The *X* field of the MATCH block specifies the number of another MATCH block, termed 'the conjugate MATCH block'. If the matching of the two events is accomplished, the first member of the pair, which was placed in an indefinitely delayed state, is made ready to advance. If its action time is zero, the transaction is placed in the current event list. It should be noted that in either case the program continues processing the transaction that has just entered the conjugate MATCH block.

gate (facility)

The GATE block is used to test the status of a simulated logical condition or an item of equipment. It is similar to the IF statement in FORTRAN. The GATE block operates by refusing entry to transactions unless the specified test is passed. It is possible to test twelve conditions by a GATE block, such as facility in use (U), facility not in use (NU), facility interrupted (I), facility not interrupted (NI), etc.

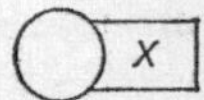

terminate

The TERMINATE block removes transactions from the system. Upon arrival of a transaction, the count of the total number of transactions that have entered the TERMINATE block is incremented by 1. The TERMINATE block is also useful in terminating a *run*, and there are two ways to do this, involving the use of TERMINATE blocks and START cards. (a) The run may be ended on a count of a specific number of terminations, coded in the *X* field of the START card. All TERMINATE blocks having an *R* coded in their *X* field will contribute towards the end count; which is kept in common. (b) The run may be ended on a predetermined value of the relative block time, coded in the *Z* field of the start card. If this method is used, no TERMINATE /*R* blocks should be in the model. Both methods of ending a run can be used simultaneously by coding both the *X* and *Z* fields of a START card and using at least one TERMINATE/*R* block. The run will then end on whichever end condition is met first.

With this background we are now in a position to draw the block diagram of the manufacturing problem. Referring to Figure 4, notice that facility no. 1 is seized at the beginning and is released only upon completion of all three units of the assembly at block 14. The cascade of SPLIT blocks generate four identical transactions, three of which represent the parts to be manufactured, and they go to ADVANCE blocks simulating the manufacturing processes. The fourth transaction occupies three units in store no. 1.

The matching operation is simply a device to guarantee seizure of the facility no. 1 until all three parts are completed. Upon entering store no. 1 (block 6), the transaction immediately advances to the MATCH block 10 where it waits. As soon as the first part is finished, it enters the conjugate MATCH block 11, and a match is realized. As a consequence, the conjugate transaction in block 11 proceeds to LEAVE block 16, where it reduces the number of units in store no. 1 by unity. At the same time, the original transaction in block 10 proceeds to ADVANCE block 12. As it attempts to exit from ADVANCE block 12, it enters a

GATE block 13 which tests for 'store empty' condition. Since only one part of the assembly has been completed, store no. 1 still has two units in 'storage', and the GATE condition is not satisfied. The program is then returned to MATCH block 10 where the cycle is repeated. When all three parts are completed, the store will become empty and the GATE condition satisfied. The assembly operation is simulated by RELEASE block 14, wherefrom the job is removed from the system. In this particular example, simulation can be set up to terminate after 100 such assemblies.

The preceding discussion of G.P.S.S. should serve to illustrate the structure and use of simulation languages.

At the time of this writing, active research is being undertaken at several places for the construction of even simpler languages which require the minimum familiarity with the computer or its language. Notable among these endeavors is an attempt to obtain structure through questionnaire (Markowitz, Ginsberg and Oldfather, 1965). Specifically, the engineer or researcher is given a set of, say, 120 questions, the answers to which are fed into the computer and serve to trigger off a compiler, resulting in a simulation program.

While these attempts are still in the embryo stage, they undoubtedly give a glimpse into the future of languages for communication with computers.

We now turn to a brief discussion of the problems of simulation.

Some problems in simulation

The idea of simulation is so patently simple that it is not easy to restrain oneself from adopting the simulation approach in almost every problem one encounters. True, it *is* a brute force approach, a fact which may be repulsive to the aesthetic nature of a few systems engineers. But they are a minority – the overwhelming majority are interested in getting an answer to their immediate problems, and the end justifies the means!

However, it is precisely these desired ends that should be uppermost in one's mind; are they really achieved by simulation? The disenchantment with the technique may be summarized in the words of Conway (1963, pp. 47–61) who has had eight years of more or less continuous use of simulation on a variety of problems and frequent discussions with other practitioners: 'I believe that, in general, simulation models take longer to construct, require much more computer time to execute, and yield much less information than their authors expected'.

But *why* is this the case? The answer lies in the fact that while the *idea* of simulation is very simple, the actual execution of the idea poses many questions of a delicate and subtle nature. While the complete understanding of these problems can be acquired only after intimate and prolonged exposure to simulation studies, a fairly good idea may be gained from a brief exposition of their nature. Unfortunately, simulation theory still lacks the definitive answers to these questions. It is hoped that as more research is done on the technique of simulation, independent of the real system being simulated, the answers will be forthcoming.

The steady-state problem. In many system simulations, it is meaningful to study only the behavior of the system under 'normal' operating conditions. Translated, this implies that all 'pipelines' are filled and that the statistical distribution of measured entities have reached their 'steady-state' distributions.

There are two ways to achieve these conditions. The first is to start the simulation with the system 'empty' and allow a long enough time of operation for all pipelines to fill and such steady state to be reached. The second is to start the system at some partially or totally filled stage and allow a shorter time of operation to arrive at the steady state.

The difficulty with both approaches is that 'steady state' is not as well-defined a 'state' as the name may suggest. It is not a characteristic of a *condition* of the system as indicated by the various measured parameters, but rather it is a property of the *long run averages* of these parameters and their statistical distributions. The only way to determine whether the steady state has been reached or not is to allow a long enough time of operation, compute some statistics, and compare these with the steady-state statistics. But how long is 'enough time?' And what about systems whose steady-state statistics are not known *a priori*, such as in simulation of nonexisting systems?

Some of the presteady-state statistics may be useful in the

study of transients in the system, but otherwise they are of little use. In effect, they represent lost time and expense. If this is combined with the uncertainty concerning being in the steady state, the analyst is justified in feeling uneasy about his conclusions. This may account for the fact that almost all reports on simulation studies preface their conclusions with the admonition that they are 'tentative and are based on a limited amount of observations', no matter how long the simulation has actually been run!

The time synchronization problem. Simulation can be conducted in real time. However, this is rarely done. Usually simulation is run in pseudotime, i.e. *machine time* which is either faster or slower than real time operation. Thus, we can imagine an internal *master clock* which is set at time zero when simulation starts and is stopped only at the end of the run. This master clock measures the internal time of the simulation. Incidentally, the ratio of real time to its representative simulation time is called the *simulation ratio.*

For example, the inventory operations of the firm over several years can be simulated in *one hour* of computer time. The master clock would simulate the passage of days, months and years, and the advance of one time-unit on the master clock may represent the passage of one day in real time. This is an example of simulation which is faster than the real system. On the other hand, the movement of a neutron through a gaseous chamber may consume, in reality, one millionth of a second but may be simulated in a few minutes of computer time. This would be an example of simulation slower than real life.

In large system simulations there are usually a large number of events that occur over time. Whenever an event occurs, the system is 'updated', which implies that the occurrence is recorded in the appropriate files, and all the necessary calculations based on it are executed. For example, in an inventory simulation, an event may be the occurrence of a withdrawal. This affects the 'on-hand' inventory file, the withdrawals statistics, and the cost statistics, as well as the ordering and replenishment procedures. In this instance, updating the system means the complete evaluation of the effects of such an event on the components of the system.

One way to keep track, i.e. synchronize, the occurrence of such events is to advance the master clock by one time-unit at a time and scan all entities and their measurable attributes (to use the parlance of SIMSCRIPT) for the occurrence of an event. If an event occurs, the system is updated; otherwise the master clock is advanced one more time-unit and the process repeated. It is easy to see that if there are k such attributes to be checked, and if the total running time is T, there will be Tk scannings in the course of the simulation. If the average elapsed (simulation) time between the occurrence of events is m, there will be, on the average, Tk/m such updatings. If m is large, many of the scannings will be fruitless in the sense that they will result in no events.

Another way to achieve synchronization is to attach to each attribute a small 'clock', henceforth referred to as the *attribute clock*. At the occurrence of an event the attribute clock is advanced to the time of the next occurrence. Now, if the attribute clocks are ranked in order of magnitude, the smallest among them denotes the (simulation) time of the next occurrence. The master clock is then advanced *directly* to that time, the event is noted, and the system updated. The process is then repeated with the newly-advanced attribute clock. It is easy to see that if k is large and m small, the frequency of occurrence of events would be so high that there is a good probability that the minimum attribute clock will cause the master clock to advance only one time-unit after all, which reduces the present approach to the previous one!

In general, if we assume the same average number of occurrences as before, Yk/m, and we know that it takes $k - 1$ comparisons to find the minimum attribute clock, the choice between the two approaches reduces to picking the minimum of Tk and $Tk(k - 1)/m$ comparisons.

Of course, better search and comparison schemes than the above can be devised. For example, suppose that the set of attribute clocks is divided into two equal subsets, I and II, and that the minimum of I is smaller than the minimum of II. Then clearly, if the system is updated and the minimum attribute clock of I advanced to the time of the next occurrence, the ranking of the attribute clocks in II will not be disturbed. Consequently, we rank only *half* the attribute clocks, namely those in subset I.

We have thus reduced the chore of ranking to $\frac{1}{2}k - 1$ comparisons instead of $k - 1$. Now, the choice is between Tk and $(Tk/m)(k/2 - 1) + 1\} = Tk^2/2m$.

Needless to say, the process of subdividing the original set of attribute clocks into smaller subsets can be carried to r subsets, $r = 3, 4, \ldots$, with the choice being between Tk and

$$\frac{Tk\{(k/r - 1) + (r - 1)\}}{m} = \frac{Tk(k/r + r - 2)}{m}.$$

However, the programming and record-keeping effort may outweigh the advantages obtained for large r.

One final remark before proceeding to a different problem in Simulation Theory. In the above discussion it was not our intention to instruct the reader on how synchronization is accomplished, since the above approaches are almost self-evident. It *is* our intention, however, to indicate a *type* of problem encountered in the construction of the simulation itself and point out the need for its consideration. Computer time wasted in search may be greatly reduced by the judicious choice of the technique used for effecting such search.

A measurement problem. In general, there are two kinds of entities: *permanent* and *temporary*. A permanent entity exists for the duration of the simulation; a temporary entity exists only for part of the simulation. For example, in air traffic simulation the runway is a permanent entity, while an airplane is a temporary entity. Entities have attributes which are measurable – for example, the runway is either occupied or vacant – and it is the changes in these attributes that constitute 'events'.

Temporary entities give rise to a peculiar type of problem in some simulation studies, in particular those involving stochastic variation. To wit, temporary entities may be created in a sequence different from that of their retirement. For instance, airplanes may 'arrive' in one pattern but may 'land' in a completely different pattern due to differences in fuel supply, mechanical condition, weather conditions, etc. The question is *which* temporary entity should be included in the simulation run?

The answer to this question clearly depends on the real system being simulated. Consider the following two alternatives: (1) Fix the duration of the run and include entities whose *complete life span* falls within the duration. (2) Fix the starting time and the number of entities to be studied and run the simulation until all these entities are created and retired.

The first procedure leaves the *sample size* as a random variable. Some entities may be created during the period of simulation but are not yet retired when simulation is stopped, and conversely, some of the 'pipeline' entities may retire during the period of simulation but their starting time may be earlier than that of the simulation. Consequently, the sample size on which deductions are to be made may be very small indeed (unless the simulation time is long).

The second procedure leaves the *duration* of simulation as a random variable. This might be undesirable, especially if a great number of runs are to be conducted, which usually is the case.

If T, the duration of simulation decided upon *a priori* in the first procedure, is large enough, the first procedure will be indistinguishable from the second as far as the sample size is concerned, though it may differ in other respects (such as initial conditions, the achievement of steady state, etc.).

Clearly, in a system where the duration is limited, such as in air attack, the first procedure is more representative of the real system. On the other hand, the inventory of a firm which does not plan to go out of business in the foreseeable future is better represented by the second procedure.

Again, it is important to realize that the decision concerning the manner in which temporary entities are included in the sample affects the very results obtained from subsequent analysis. It is possible that a comparative advantage of one policy over another is obscured or even reversed simply by a change in the method of determining which temporary entities are to be included in the sample.

The question of validity. For the purposes of our present discussion, validity is to mean correct representation. But representation is mostly a subjective feeling, because 'likeness' is hard to assess. One can, of course, demand as a minimum prerequisite for validity that for at least one set of conditions the simulation should produce results consistent with the performance of the real system. Yet this is no *guarantee* that the simulation is a valid representation, since it may fail the test under a different set of conditions. Such a test is essentially a null test of hypothesis; if the test fails, the simulation is firmly declared *invalid*, but if it succeeds no such firm assertion can be made concerning its *validity*. On the other hand, one cannot demand that the simulation yields consistent results under *all* conditions, for then there would be no need for simulation in the first place!

The result of simulation is a measurement on an attribute of a permanent or temporary entity. In Monte Carlo simulation, this measurement is subject to error. The validity of the measurement is then dependent on the magnitude of the variability of the measurement, that is, its *precision*. The question of validity therefore resolves into the question of precision in the statistical sense.

Randomness is inherent in physical experimentation, even in completely deterministic systems, due to the random error in the measuring devices. The magnitude of such error is known in digital simulation since it is the error in the rightmost digit of the measurement, and this can be made as small as desired. The second and most important source of randomness in simulation is the variation in the exogenous input data. This, in turn, can be controlled in two ways. First, a fixed set of real-life input data is used on each alternative policy under investigation and the different responses to this one set compared. For example, if a new rule for inventory replenishment is suggested, it can be tried on actual demand over the past five years. The result of the simulation is compared with the performance of the real system under the existing rules, and the better rule is chosen. The major argument against such a procedure is that the performance of the new rule has been tested under *one* set of conditions which may be, by chance, particularly favorable or unfavorable to it.

The second approach is to generate a set of random demand drawn from a distribution possessing the same statistical characteristics as actual demand. Although a certain appearance of impartiality has been achieved, the same objection still holds.

We seem to have reached an impasse. It is generally acknowledged that the best way to compare two alternatives is 'under identical conditions'. But whenever that is achieved, the result is open to dismissal as being not necessarily representative!

All arguments seem to indicate that 'identical conditions' should be taken in a statistical sense, and that alternatives must be compared by repeated experimentation under a stationary *causation system*. Replication through choice of different sets of random numbers guarantees independence of measurements but requires a prodigious number of experiments. Moreover, whenever possible, some 'scaling' of the simulation results should be made in relation to the real system.

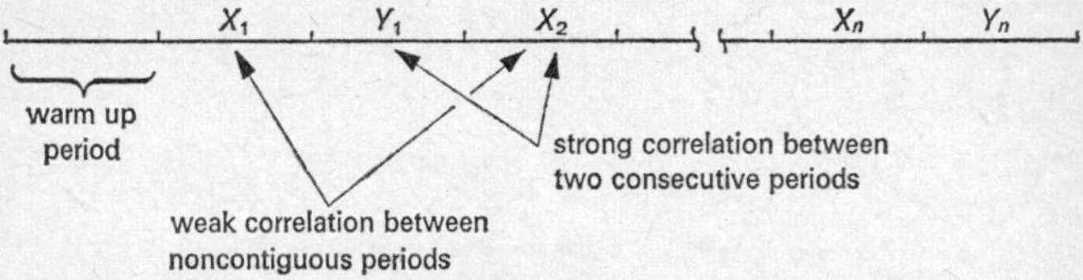

Figure 5

Precision can be measured by the *variance* of the experimental results, and the variance can be minimized by averaging over a large sample size. Let x_i denote one such average; for example, x_i may be the average inventory on hand over the simulated period. The variability of X_i (capital letters denote the random variable and small letters the value they assume *after* the experiment) can be determined through replication, say n times. Each replication requires a 'warm up period' to achieve steady-state conditions, which is wasteful in time and effort. It can easily be shown that a long run of size $2n + 1$ experiments (the one experiment is discarded for 'warming up') yields a global average which is 'better', that is, more precise, than the average $\bar{X}$ of the separate n experiments. Variability of this global average, denoted by $\bar{Z}$ can be obtained by 'chopping up' the long run into $2n$ sets: the X_is and the Y_is, which are alternately located, as shown in Figure 5. For let $\bar{X}$ and $\bar{Y}$ denote the averages of the two subsets, respectively, then

$$\bar{Z} = \frac{\bar{X} + \bar{Y}}{2}$$

whose expected value is equal to the expected value of $\bar{X}$ on the n experiments. But comparing the variances we have

$$\text{var}(\bar{X}) = \frac{\sigma^2}{n} = \text{var}(\bar{Y})$$

while

$$\text{var}(\bar{Z}) = \text{var}\left[\frac{1}{2n}\sum_{1}^{2n} X_i\right] = \frac{1}{4n^2}[2n\sigma^2 + 2(2n-1)\,\text{cov}(X_i, X_j)]$$

where $\text{cov}(X_i, X_j)$ is the covariance of two separate X_is. Since the determinant of the covariance matrix is always nonnegative, and in this case the covariance of all nonadjacent variables is assumed to be zero, it follows that $\text{cov}(X_i, X_j) \leqslant \frac{1}{2}\sigma^2$. Consequently,

$$\text{var}(\bar{Z}) = \frac{1}{4n^2}[2n\sigma^2 + 2(2n-1)\,\text{cov}(X_i, X_j)]$$

$$\leqslant \frac{\sigma^2}{2n}\left(1 + \frac{2n-1}{2n}\right) < \frac{\sigma^2}{n} = \text{var}(\bar{X}).$$

When the correlation of nonadjacent values are significant, the variance of $\bar{Z}$ is given by

$$\text{var}(\bar{Z}) = \frac{\sigma^2}{4n^2}\left[2n + 2\sum_{s=1}^{2n-1}(2n-s)\,\rho_s\right],$$

where s is the spacing between two X_is and ρ measures the correlation between X_i and X_{i+s}. This correlation is usually significant and should be estimated from pilot runs to determine how measurements are to be taken. Moreover, they should be checked again during the actual experimentation.

Unfortunately, the problem of validity is complicated by the fact that more than one statistic is usually derived from the same set of data. These statistics are then utilized in several calculations, and because of the strong correlation among these statistics, it is extremely difficult to assess the variability of the results.

In view of the above discussion, the reader may wonder how *any* conclusions can be drawn from simulation. Fortunately, simulation is used most often to compare alternatives, and *absolute* measurements are not important. What is important is the *relative* preference of one alternative over another. Even under highly imprecise measurements, some idea can be gained about such relative advantages.

A final question connected with the study of validity is the question of *initial* or *starting conditions*. We have seen above how the starting conditions may affect the time of reaching the steady state. Here, we wish to discuss the effect of the starting conditions on the measured result itself.

Suppose that simulation is used to compare two alternatives, *A* and *B*, and suppose that enough is known about each alternative to realize that there is an appreciable difference between them under steady-state operation. Each alternative possesses, however, its own 'reasonable' starting conditions. The question is to determine the magnitude of that difference. Is it justifiable to start each alternative under its own advantageous starting conditions, or should a common starting condition be set for both alternatives?

The answer to this question is not easily determined. It seems 'unfair' to suspect that project *A* is better than project *B*, to start both projects under their most suitable conditions, obtain a result which confirms the suspicion, and then use this result as a 'proof' of the superiority of *A* over *B*! And yet it is equally 'unfair' to handicap *A* by any adverse starting conditions if, in fact, it can be started in real life under favorable conditions and continue to operate under equally favorable conditions!

Perhaps a possible way out of this dilemma is to measure the performance of both projects under identical starting conditions which are a compromise between the two conditions favorable to either project individually *as well as* measure the performance of both projects under their respective most favorable conditions. The cost of added experimentation may be amply justified by the insight gained concerning the relative performance of the two projects.

Simulation as a general problem solving technique. Several other problems of simulation exist, the majority of which are peculiar to the simulated system and therefore vary from problem to problem. For example, in systems of several concurrent inputs and events the problem of synchronization takes on more serious dimensions. In systems possessing variable environments, the problem of adaptability of the simulation to change in its very structure becomes important, and so on.

Simulation as a problem-solving tool is still a novel development. The above brief list of problems should caution the analyst against setting high hopes and glossing over important details. In particular, the following are some of the reasons given *against* simulation:

(a) It is a slow and expensive undertaking. In the majority of cases it requires large-scale computing facilities which may not be available. A simulation may take several man-years to construct and several more to run and analyse the results, with no guarantee of optimality.
(b) It is imprecise, and there is no known measure of the degree of its imprecision.
(c) It is almost always a specific tool for a specific application, in spite of the large expenditures in effort, time, and money that usually go into it. One rarely encounters a simulation which has been used in more than one study.
(d) Because the output is in numerical terms, a certain degree of definitiveness is accorded to the results and more is read into them than is really justified. In many cases, the underlying imprecision renders the whole mass of data meaningless.
(e) Simulations have a tendency to grow large in spite of the analyst's conscientious effort to keep their size to a minimum. In fact, a simulation is never complete; there are always a few more experiments that should be conducted to render the picture clearer

Examples of the use of simulation as an aid to design

It is safe to state that the past few years have witnessed the use of simulation as a design tool in an ever-increasing variety of undertakings in widely varied fields. This is attested to by the number of symposia that have been held (and the number of books and articles that have been published) concerning the extent to which the technique of simulation has permeated the various fields of science and engineering as a basic tool of design. It is not our intention (nor is this the place) to present another survey of the same area. Suffice it here to mention that, apart from classical

Monte Carlo applications to engineering and mathematical problems, simulation has been extensively used in both the military and civilian systems to the point where, today, it is almost impossible to find a problem in either field which has not been treated in one formulation or another by the technique of simulation. Morgenthaler (1961, pp. 392–405) gives a fairly complete survey of the various problems attacked by this approach, and lists ninety-three references on the subject of applications of simulation and gaming. In this section, to illustrate the use of simulation as a tool of design, we pick up the thread of discussion of a problem for which analytical models have been proposed and pursue the matter further. In this fashion, we are, hopefully, able to impress upon the reader that simulation is *another approach* to the same problem.

Example: the warehouse location–allocation problem

The general problem of warehouse location–allocation can be stated as follows: given a set of factories and customers, and given all relevant information about them, it is required to determine:

how many warehouses where are the warehouses to be located	Location problem
which customers to be served by which warehouse what volume should each warehouse handle what modes of distribution should be used	Allocation problem

and minimize some overall cost function.

The problem in its general form has defied complete analytical solution because of its complexity. Various factors contribute to such complexity:

1. Variation in *customer characteristics* relative to: size of order, frequency of ordering, mode of transportation (rail, ship, truck, air) and product mix.

2. Variation in *factory characteristics* relative to: product mix, capacity of each product line, and the cost of changing this capacity.

3. Variation in *warehouse characteristics* relative to: the service

policy as indicated by the desired reliability of satisfying demand and by the acceptable delay in delivery, together with the functional relationship between such delay and customer demand.

4. *Nonlinearities* in the overall cost function: the transportation costs are, in reality, nonlinear from factory to warehouse and from warehouse to customer. The cost of transportation varies with the size of the order shipped, the distance it travels, the geographic location of roads, and the means of transportation. Moreover, warehousing costs are nonlinear, the overhead costs are of the 0, 1 type, since they are incurred if the warehouse exists and they are not if the warehouse does not exist. The variable costs are nonlinear functions of the size of the throughput and the required safety stock (usually taken as proportional to the square root of the throughput).

Several approaches have been proposed for the solution of the problem. They can be classified under two categories: mathematical approaches and trial and error approaches. The first category depends on constructing a mathematical model, which embodies many simplifying assumptions, and attempts to arrive at an 'optimal' solution.

The second category tries several 'reasonable' solutions and picks the best from among them, but it makes no claim to optimality. In this section, we discuss a *simulation* approach.

A study of both categories, mathematical as well as trial and error, reveals that none of the attacks on the problem necessarily achieved optimality, and that each investigation attempted to break up the large problem into two smaller ones. The consensus seems to be to determine the *location* of the warehouses in some heuristic fashion and then solve the allocation problem optimally. The rationale behind such an approach is that the choice of location is usually constrained by many external factors which drastically reduce the theoretically possible alternatives to a relative few. The determination of the optimal allocation for each alternative location pattern is then a much easier task.

The approach using simulation. Rather than talk in generalities, we consider a concrete case. This is an application which was conducted by Shycon and Maffei (1960), for the H. J. Heinz

Company, a large food products manufacturer, in which the simulation approach was used.

Typically, production is performed at a few factories in different locations (and of different capacities and products) and the product is shipped to a number of warehouses strategically located throughout the continental United States. Certain factors forced the company to take a fresh look at its distribution system. These were primarily related to shifts in population centers and principal markets, the emergence of brand identification as a prime marketing factor, technological changes in distribution methods, the growth of a large retail operations, and the changes in marketing techniques. Added to these, of course, is the fact that the cost of the physical distribution of products to markets has been rapidly increasing.

A little thought reveals that if all customers were large consumers, if all of them gave the factories sufficient advance notice (i.e. lead time) of their requirements, and if all factories produced the full line of products, there would really be no need for warehousing. Warehouses exist because the three 'ifs' listed above are not realized. The problem confronting the company then is the following: how to design its distribution system with *production* carried out at the factory, *mixing* (of different products to satisfy a given customer's demand) carried out at 'mixing points' which sometimes coincided with the factories and sometimes with the warehouses, and *storing* carried out at the warehouses.

The characteristics of each component of the system were considered in great detail. A partial list of these characteristics has been presented above (p. 355). A simulation was constructed which permitted the study of a large number of warehouses and thousands of demand centers in different geographic locations. Any alternative configuration was represented by a particular location of the warehouses and a particular allocation of distribution. Simulation by Monte Carlo methods under any given configuration resulted in a vector of values of certain measures of performance which are of primary interest to management, such as the cost of transportation, cost of warehouse operation, delay in delivery, level of customer service, etc. By varying the configuration, a new vector of values was obtained, and repeated experimentation led to the decision on the 'best' configuration among those considered.

As was pointed out above, the basic impetus to such simulation was to evaluate alternative policies of distribution. But once the simulation was constructed, it became obvious that it could be used with equal facility to evaluate policies concerning other aspects of the company's operation. This is typical of the simulation approach; the very act of constructing a simulation uncovers underlying assumptions, previously determined practices, and alternative ways of performing old or contemplated functions, which led to new avenues of research not contemplated at the beginning of constructing the simulation. In the words of Shycon and Maffei (1960, p. 74):

The method for performing the Heinz study has provided great facility for studying other aspects of the business which are at least as important as the development of an optimal distribution system. When the simulation was developed, it was thought important to build a general-purpose tool, one which management could use at any time, future as well as present, to study questions of major concern. It was not, however, until the simulation was designed that it was fully realized that the tool provided such facility for studying a wide range of perplexing management problems. Specifically:

Distribution cost studies. Customers can be separated by areas, types, shipment sizes, salesmen, type of carrier, channels of distribution. We could get estimates of distribution costs on the basis of each or any combination thereof.

Locational studies. The number and location of factories could be changed, for example, rather than altering the warehouse configuration. Then, too, the effect on the company's operations of a sudden shift in customer type or location could be studied.

Studies related to products. The product mix at each factory can be changed arbitrarily to observe whether adding product capacity would change distribution costs appreciably. Similarly, customer consumption patterns can be altered to see what effect such changes will have on distribution costs.

Studies related to time. Customer data can be altered in order to reflect gross annual volume changes by product line. These data would then be used to determine distribution costs. Thus, it can be seen what effect proposed changes in sales policy, prices, or new products would have on customer purchasing frequency, order size or volume. The possible effect on distribution costs and on profitability can be estimated experimentally.

As a footnote to the various policies considered, any customer who can be satisfied from any factory was so designated, and that portion of designating a demand level at which it is economical to ship directly to the customer. The remainder of the demand was the subject of investigation through simulation.

What are the advantages of such an extensive and expensive undertaking? First, the achievement of a degree of realism not achieved by any other approach. Second, the availability of a tool of analysis for distribution *as well as* manufacturing and marketing policies. If the tool is utilized as extensively as its capabilities indicate, there is little doubt that it more than pays for itself.

References

CONWAY, R. W. (1963), 'Some tactical problems in digital simulation', *Management Science*, vol. 10, no. 1.

GORDON, G. (1962), 'A general purpose systems simulator', *I.B.M. Systems J.*, vol. 1.

KAVANAUGH, T. F. (1961), 'TABSOL, the language of decision making', *Computers and Automation*, vol. 10.

MATHEMATICS AND APPLICATIONS DEPARTMENT, I.B.M. (1960), 'The job shop simulation', I.B.M. Corporation, Data System Divisions, New York.

MARKOWITZ, H., GINSBERG, A. S., and OLDFATHER, P. M. (1965), 'Programming by questionnaire', RAND Memorandum RM-4460-PR.

MORGENTHALER, G. W. (1961), 'The use and application of simulation in O.R.', in R. L. Ackoff (ed.), *Progress in Operations Research*, Wiley.

SHYCON, H. N., and MAFFEI, R. B. (1960), 'Simulation – tool in better distribution', *Harvard bus. Rev.*, vol. 38, no. 6, pp. 65–75.

21 D. F. Boyd

Enterprise Models: A New Management Technique

D. F. Boyd, 'Enterprise models: a new management technique', *Industrial Management Review*, vol. 8 (1966), pp. 55–70.

In April of this year, a story appeared in the business section of the *New York Herald Tribune* that was perhaps more significant than it appeared on casual reading. The story was about a computer-based simulation model being developed for Moore-McCormack Lines, Inc., for use as a 'tool for short- and long-range planning'. What was not evident from the story was that Moore-McCormack is developing not just another computer-simulation model but rather a new class of models, called enterprise models.

The distinction between conventional computer models and enterprise models is not so much a matter of technique but of viewpoint and application. Both use standard computer-simulation techniques to represent an operation. But, where most conventional models focus on measures of physical activity and/or their related costs, enterprise models extend the representation to the level of profit centers and emphasize overall economic evaluations of primary interest to upper levels of management. The difference is highly significant.

Models limited to a cost center tend to concentrate on representing the men, materials, equipment, and procedures that make up a physical operation. Little, if any, of the financial and management-control framework that ties the modelled operation to the rest of the organization is represented. The result is a model of the operation from the viewpoint of supervisory management. The utility of such a model is limited to showing the effects of physical changes, such as queueing phenomena, equipment capacity, layout, and work flow, and these effects are shown in terms of physical parameters, such as equipment utilization, lead time, delivery performance, or throughput, which are meaningful principally at the supervisory-management level.

The scope of enterprise models is considerably broader. They may represent a profit center comprised of several cost centers. Consequently, they show not one but several physical systems as a single, integrated operation. In addition, they include the management-control and financial framework that integrates the individual cost centers into the overall organization. The result is a model of an operation as seen from a level of management at least one step higher than that responsible for a single cost center.

The utility of this type of model is both different from and broader in scope than the cost-center model. First, the difference: In general it will contain less physical detail in its representation of operations than is possible with a cost-center model. However, it can be used for higher-level planning by evaluating the effect that changes in individual cost centers will have upon all other operations in the profit center.

Even more important, the enterprise model can be used also to show the effects of changes in management controls and financial procedures and policies. It can provide useful answers to a wide range of 'what if?' questions that could be answered formerly only by trial or intuition. Would direct costing be a more effective management tool than standard cost accounting? At what levels of backlog and demand should an extra shift be added or overtime authorized? What is the real cost of handling super-priority orders? Should a plant produce solely to stock or solely to order? Or should production policy be set somewhere between these two extremes?

The answers that an enterprise model supplies are consistent with the management viewpoint it reflects. Effects are shown not only in physical rates and levels, but in financial terms as well. The financial-performance data are used primarily to identify and to analyse the reasons for the financial-performance levels.

Management Orientation

The unequivocal management orientation of enterprise models makes them a very promising tool for helping management in making a wide range of operational, tactical, and strategic decisions. In this respect they may be compared to certain formal analytical mathematical models and optimizing techniques of demonstrated value in special cases but whose application has been limited by two factors. For one thing, they can be usefully applied only to situations for which there is a 'general theory' that can be expressed in precise mathematical terms or for which narrow assumptions as to linearity or equilibrium may be acceptable. Most management problems don't fit into this category. A further shortcoming of mathematical techniques is their high level of abstraction. Mathematics is an abstract language not universally understood by managers, who often find it difficult to translate mathematical abstractions into real-world terms. Consequently, a large communications gap may exist between such models and the management user.

Enterprise models ease both of these problems. A comprehensive understanding of the total dynamics of the operation being studied is not needed before the model is begun and, indeed, is not likely to exist. The model may be built a piece at a time, and test runs provide insights to how the pieces interact. The building process itself helps fill in the gaps of knowledge about how an operation works. This makes enterprise models useful in many areas in which mathematical models are not applicable.

Enterprise models aid communications by depicting a situation in terms familiar to management. Although few managers have the training to construct the model, all of them can understand the flow charts of a model well enough to verify its structure and contribute to its design.

Another advantage of enterprise models is that, unlike most formal mathematical models, they are open-ended. They can easily be expanded to include additional functions and relationships. A model built for operating management of a manufacturing plant can be extended over a period of time to reflect the changing needs of its original user or to include the viewpoints of higher levels of management. This can be done without completely reformulating the original framework of the model. This continuity of use gives enterprise models a broader base of economic justification than specialized, single-purpose computer-simulation models have had in the past.

Anatomy of an Enterprise Model

An enterprise model structures an operation as a servo-mechanism of the type shown in Figure 1. A decision center (management) regulates the operation of a physical process. The planning and control decisions that control the process are

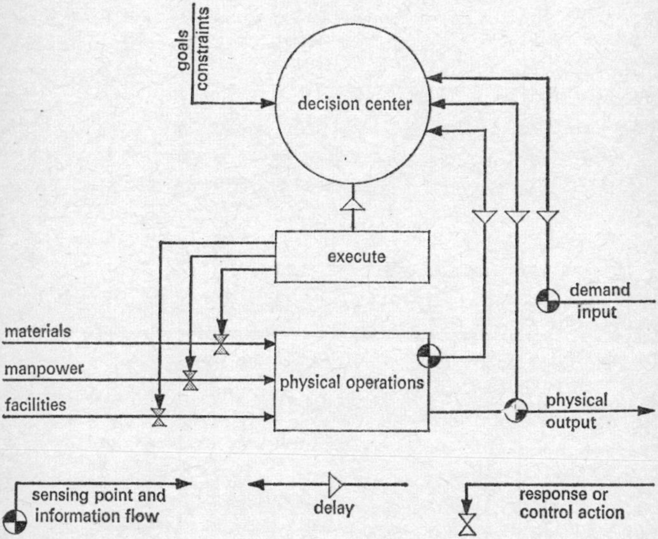

Figure 1 Conceptual framework for the basic response mechanism of an enterprise

based on process performance information fed back to the decision center and on goals and constraints imposed from the next higher level of management.

This concept can be extended beyond the operational management directly concerned with the physical process to the higher levels of tactical and strategic management. Such an extension is shown in Figure 2.

An enterprise model can be designed to emphasize any level of this management hierarchy. A typical subject is represented by four networks. One shows the physical process in terms of operations and the flow of materials. The second network depicts the flow of production information, such as customer orders,

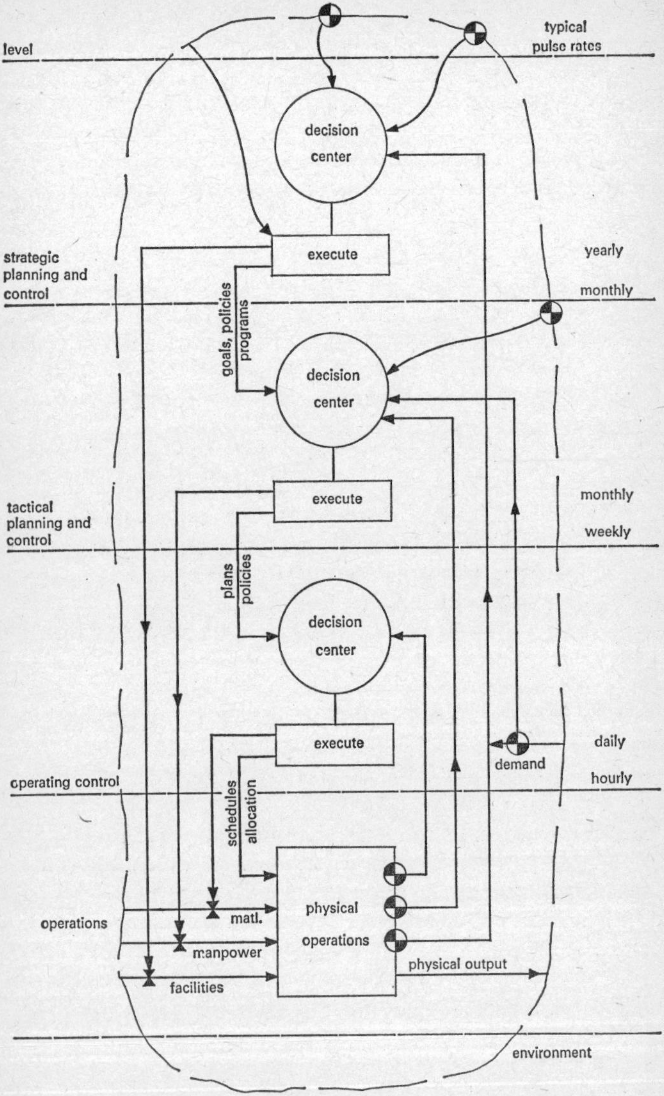

Figure 2 Hierarchy of planning and control levels in a corporate enterprise

which establish the demand placed on the physical process. The third network depicts the flow of planning and control information, which determines how the physical process will respond. This third network includes channels for reporting the performance of the physical process and decision rules that reflect management policy. The fourth network depicts the accounting system and financial framework which interpret the physical activity in economic terms.

The amount and type of detail these networks depict vary according to the management viewpoint of the model. Two experimental models which have been constructed may be compared to show the extent of the variation.

One of these models reflects the operating-management view of an integrated steel mill. It is focused primarily on the problems of scheduling and allocating resources within the mill. Consequently it represents the physical processing operation in considerable detail, but doesn't include much of the planning and control network that links operational management to higher management levels.

The second model is focused on the tactical and strategic management levels of an aerospace firm. It is concerned primarily with the control of overall manpower levels, the development of broad plans and policies for governing materials procurement and resource allocations, and the establishment of long-range operating and financial goals and programs. Consequently, it represents the physical processing operation only in general, aggregate terms, but shows the planning and control network at the tactical and strategic levels in considerable detail.

Although the two models are different in their management orientation, they employ the same fundamental techniques. The following description of the steel-mill model illustrates these techniques and their application.

The Steel-Mill Process

The steel-mill simulation model represents the physical process as an integrated group of six operating units:

1. Open hearth.
2. Primary mill.

3. Blooming and billet mills.
4. Hot-strip mill.
5. Hot-rooled coils finishing.
6. Cold reduction mill.

The sequence of processing operations is shown in Figure 3. Processing begins at the open hearth, which produces various grades of steel. The steel is poured into ingots which are then moved to the soaking pits associated with the primary mill. These pits act as a storage area and at the same time serve to stabilize the temperature of the ingots. The primary mill rolls the ingots into square blooms or flat slabs. Blooms are subsequently processed in the blooming and billet mills.

The slabs move to the hot-strip mill by one of three routes with the choice based primarily on the occurrence of defects. They may go directly from the primary mill. They may go through a short reheating cycle, which brings them up to the required temperature level. Or, they may go through the slab yard, where they are allowed to cool completely.

The output of the hot-strip mill may go directly to the finishing operation or to the cold reduction mill for additional processing.

The model describes these physical operations in considerable detail. It represents each of the processing steps and related storage functions, the flow of materials through the processing-storage network, the production of scrap, and the breakdown and repair of the individual operating units.

Most important from the viewpoint of operating management, it also describes the processing constraints that one operating unit imposes upon another. For example, the minimum amount of time an ingot must remain in the soaking pit is determined by the amount of cooling that has taken place during its move from the ingot casting stage. Thus, the materials-handling facilities at the open hearth affect the requirements for soaking-pit capacity and the availability of ingots for subsequent processing stages.

There is also a relationship between the primary and billet mills. Billet-mill operation requires a flow of material at suitable temperature directly from the primary mill. Consequently, these two units must operate in tandem, and their processing rates must be balanced.

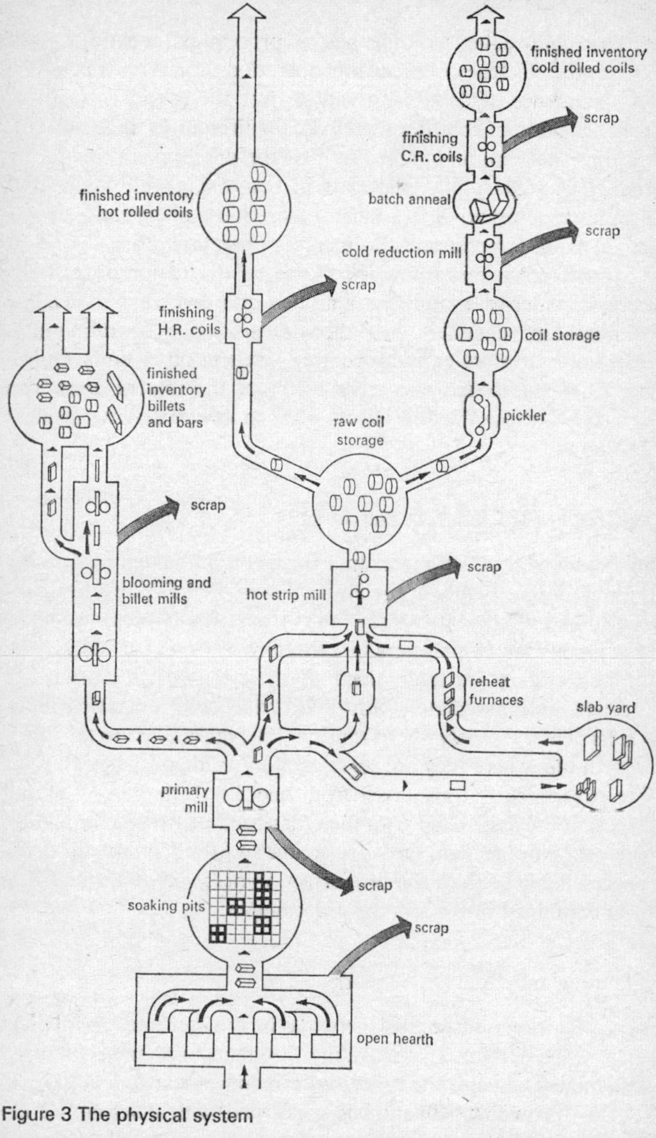

Figure 3 The physical system

The cold-reduction mill has a processing constraint that affects the efficiency of preceding operations. Coils must be rolled in a sequence of decreasing widths over the life of each set of rolls. Maximum roll life, therefore, requires either that substantial inventory be maintained or that the operations prior to the reduction mill supply materials in decreasing-width sequence. The latter alternative complicates these prior operations; on the other hand, large inventories increase operating costs.

These interrelationships show one of the major distinctions between enterprise models and models of individual cost centers. Changes in schedules and allocations in one operating unit affect utilization rates and inventory levels in other units. Only a model whose scope encompasses all of the related units, can accurately show the full effects of such operating-management decisions.

Production-Information Network

The flow of production or 'mill' orders is simulated in the steel mill model. In Figure 4, the physical system of Figure 3 has been repeated in schematic form, and orders are represented as a document flow which parallels the physical flow of steel. Files of mill orders are maintained at the open-hearth and at the storage areas associated with each of the other operating units. Orders enter the network at random intervals and go to the open-hearth file, where they are segregated by grade of material. They are subsequently transferred from one file to the next when all the material associated with them has been processed through to the next storage area. Order selection and the sequence of order processing at each of the processing points is based on information contained in the storage-order file.

Planning and Control Network

The planning and control network in the steel mill is divided into three elements: 'providing', sequencing, and operating schedule. These are the three major management functions.

The 'providing' function is concerned with determining the amount of steel the open hearth must produce to fill the order.

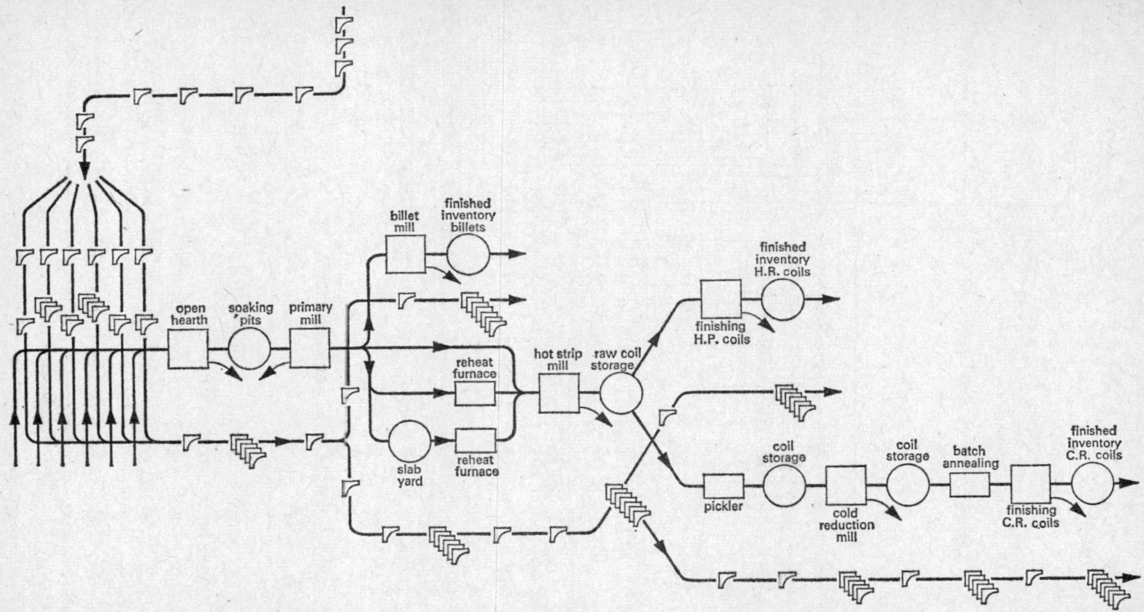

Figure 4 Flow of mill orders

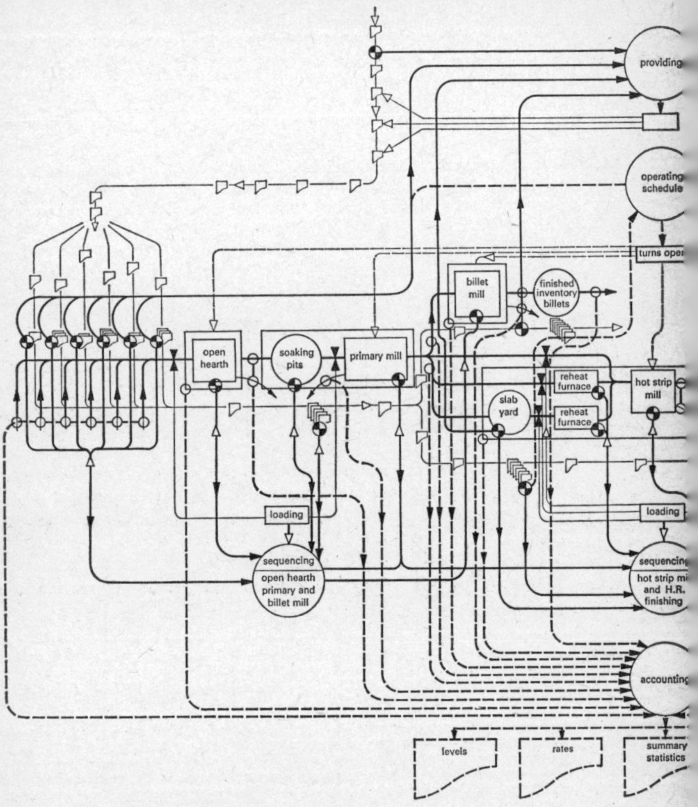

Figure 5 Schematic diagram, complete model

It is also concerned with estimating flow time through the mill and assigning a 'promised date'.

The sequencing function is concerned with determining the sequence in which orders will be processed at each of the operating units.

The operating-schedule function is concerned with specifying, by shift, which operating units will be 'up' and 'down' during the week.

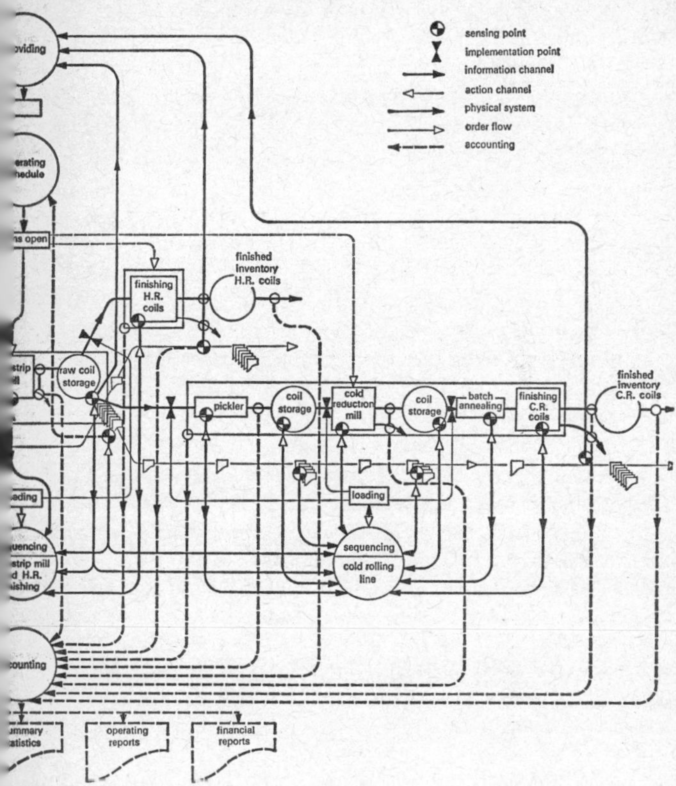

The three parts of the planning and control network are shown schematically in Figure 5. Information-collection points, decision centers, order-implementation points, and the information channels that link them all together are indicated. Two types of information channels are shown for each function. One represents the flow of information to the decision centers. The other represents the transmission of decision-center instructions to the implementation points.

The simulation model includes a set of specific rules which make the providing, sequencing, and operating-schedule decisions. One of the functions of the model is to test alternative decision rules.

The rules for 'providing' decisions call for the model to determine the expected scrap rates associated with each order and take into consideration the availability of material on hand that can be used to help fill the order.

The sequencing rules for most of the operations give preference to orders with the earliest 'promised date'. At the hot strip mill, this preference is applied within an additional priority framework. The highest priority is given to material that is acceptable for direct movement from the primary mill for processing in the hot-strip mill. Material moving out of the reheat furnace is given second priority. Materials coming from the slab yard have lowest priority. But order sequencing within these three priority levels is based upon the earliest promised date.

Although the same general sequencing rule is used for orders entering the cold-reduction mill, the orders are resequenced for part of the operation to meet the requirement for coils of decreasing widths. Material leaving the reduction mill is then sequenced by temper requirements, and coils are selected for finishing from the longest queue.

The rules for generating an operating schedule call for calculating the number of full shifts required to process the work load ahead of each operating unit. Average throughput rates are used in the calculations. The required number of shifts are scheduled for all units, taking into account interdependent units that must be scheduled in tandem.

Accounting Network

The accounting network in the model (also represented in Figure 5) is designed to report mill performance in both physical units of production and in financial terms.

Data on physical production is supplied in a weekly operating report (see Table 1). A series of special statistical tabulations may also be generated for deeper analysis. The operating report provides detailed descriptions of the order backlog at the open

Table 1

Weekly Operating Report

Week 8					
Backlog status					
Grade	*Orders*				
A	12				
B	22				
C	20				
D	2				
E	23				
F	127				
Inventory status					
Unit	*Ingots*	*Slabs*	*Coils*	*Tons*	*Orders*
Soaking pit	533				
For primary (blooms)				1764	
For primary (slabs)				7678	
Slab yard		811		8868	
For hot finish				759	
For cold finish				5816	
For cold reduction				7397	
For batch annealing					
1			68		
2			18		
3			32		
4			131		
5			158		
For temper finish					
1			0		
2			0		

hearth and the inventory at each of the other processing units. The statistical tabulations can supply detailed or summarized data on the rates and levels of processing units and stores for any reporting interval specified by the user. Statistical distributions of key operating parameters, such as transit time through the mill, also can be provided.

Table 2

Weekly Manufacturing Expense and Cost Performance Report

Week 8

Weekly manufacturing expense and cost performance report

	Open Hearth	*Primary Mill*	*Hotstrip Mill*	*Blooms Billets*	*Hot Finish*	*C.R. Line*	*Total*
	CC 1	CC 2	CC 3	CC 4	CC 5	CC 6	
Expense Statement:							
Direct material							444,360
Direct labor	11,760	7700	5250	8820	280	10,640	44,450
MFG. Expense:							
Indirect labor	1274	833	539	931	0	1127	4704
Supplies	4410	1540	2100	630	210	2660	11,550
Fuel	17,640	7700	8400	0	0	5320	39,060
Power	1470	3850	10,500	792	105	6650	23,367
Other expenses	341	223	144	249	0	301	1258
Depreciation	37,500	12,500	37,500	12,500	6250	18,750	125,000
Sub Total MFG. EXP.	62,635	26,646	59,183	15,102	6565	34,808	204,939
Total expenditures	74,395	34,346	64,433	23,922	6845	45,448	249,389
Operating rate:							
Equiv. tons processed	8820	8528	7539	1511	1228	5594	
Capacity (21 turns)	14,700	16,800	13,440	4200	8400	8400	
Per cent of capacity	60	50	56	35	14	66	
Cost performance:							
Cost per ton	8	4	8	15	5	8	
STD. Cost per ton	8	7	10	13	4	11	
Cost variance	0	—3	—2	2	1	—3	
Variance, per cent	0	—42	—20	15	25	—27	

Financial data is provided in a weekly report on manufacturing expense and cost performance and in summary financial reports. The data is generated on a cost-center basis. The accounting network divides the mill into six functional cost centers and records the units of material moving into and out of them. These figures are translated, using actual or standard cost values, into dollars.

The weekly report on manufacturing expense and cost performance (see Table 2) shows all operating expenditures, except raw materials, by cost center. It also shows utilization of equipment and a comparison between standard and actual costs by cost center.

The financial reports are generated for each simulated month and include income statements and balance sheets (see Table 3).

The income statement shows both sales and interest income or expense. Sales costs are broken into standard costs and manufacturing-cost variances. Net profit on operations and cash flow are also shown.

The balance sheet shows cash, the value of in-process and finished inventories, notes receivable or payable, and working capital at the end of the reporting period.

The interest income or expense on the income statement, and the notes receivable or payable on the balance sheet are part of a mechanism by which the model imposes a financial penalty on excessive inplant inventories. The mill begins each month with a specified cash level. If inventory investment produces a cash drawdown, funds are borrowed to bring cash back to the specified level. This produces an interest expense during the next month. Conversely, if operations produce a cash buildup, the excess is lent, and interest income is produced for the next month.

Importance of Financial Framework

The inclusion of an authentic and familiar accounting and financial network in an enterprise model is of great importance. It serves the same vital role in the simulation model as in the real world, namely that of providing a higher-order value framework within which diverse and complex phenomena can be compared, and ranked in economic terms.

Table 3

Monthly Financial Statements

Week 8			
Income Statement			
Sales			2,982,024
Interest income-expense			−5124
Total revenue			2,976,900
Deduct			
Standard cost of sales		2,849,539	
Manufacturing cost variance		−112,186	
Cost of sales			2,737,353
Gross profit on sales			239,547
Less			
Selling expense		30,000	
General and admin. expense		70,000	
Total selling and adm. expense			100,000
Net profit on operations			139,547
Cash flow			712,405
Balance sheet			
Cash		1,212,405	
In process inventories			
Open Hearth	15,120		
Soaking pits	511,377		
Billet mills	1288		
Slab yard	515,049		
Raw coil storage	426,060		
Pickled coils	403,716		
Annealing	326,656		
Total in-process		2,199,266	
Finished inventories			
Billets and bars	3040		
H. R. Coils	0		
C. R. Coils	72,156		
Total finished		75,196	
Notes receivable–payable		−1,024,987	
Working capital		2,461,880	

For example, in the case of the steel mill model, two alternative simulations based on the use of different decision rules might exhibit the following physical behavior patterns:

1. 'Low inventory levels were maintained, but facility utilization was only fair.'
2. 'Inventory levels were generally high, facility utilization was excellent.'

Which simulation represents the best mode of mill operation from the viewpoint of top management of the mill? This question simply cannot be answered in a satisfactory way based on purely physical measures of performance. The accounting and financial framework of the model compares the *profitability* of the two modes of operation and thus provides an acceptable basis for choice from a management viewpoint.

Enterprise models can provide accounting and financial output statements that are virtually identical in form and context with those used in the actual firm described by the model. In this way, results are communicated directly to management in the most natural and convenient way without the necessity for translation or complex explanation.

Model Applications

An enterprise model of the steel-mill type permits useful experiments to be carried out which relate to many of the functions and responsibilities of an operational-management staff. In fact, the model's ability to show the effect of financial and production policies on the physical process has a degree of utility at the tactical management level as well. Here are some typical applications at both levels.

Operating rules and procedures

The decision rules in the model can be replaced with alternative rules to see which produce the best results. A variety of sequencing rules, ranging from a simple first-come-first-served rule to a complex linear-programming decision-mechanism can be tested.

For example, consider two alternative sequencing rules for processing orders through the primary mill. The computer program logic for each of the two rules when converted to a statement in English might appear as follows:

Rule 1: Scan the file of mill orders covering all ingots in the soaking pits at the beginning of the shift. Sort these by grade of steel, ignoring the 'promised date' and roll all ingots of a single grade in succession before changing to the next grade.

Rule 2: Scan the file of mill orders covering all ingots in the soaking pits at the beginning of the shift. Sort these orders on the basis of 'promised date', ignoring grade, and roll all ingots in the sequence of earliest promised date.

The implementation of each of the above rules will have many subtle and dynamic effects on the entire mill. Rule 1 should result in some improvement in throughput of the primary mill due to fewer changes in the special settings needed for each grade. On the other hand, by ignoring promised dates some orders may be subject to late delivery.

Rule 2 should improve deliveries on or before a promised date but at the expense of reduced throughput of the primary mill.

The ultimate effects of the two different rules will also depend on the sequencing rules being used in subsequent processing units. As noted earlier, the model is capable of evaluating these complex interactions not only in physical terms but also in meaningful economic terms through the accounting and financial framework.

The frequency with which decision rules should be applied can also be tested. For example, the model can show the effect of resequencing orders every hour instead of once per shift.

Information system design

Important aspects of information systems can be tested while they are still at the functional-design level, before they are implemented. The model makes no distinction between manual or automated information flows or between human or machine decisions. Consequently, useful evaluations of a proposed information system can be obtained before detailed decisions have been made about the means for implementation.

Plant and facility design

Although the simulation model doesn't represent the physical process in sufficient detail to help with the detailed design of individual operating units, it is useful in evaluating design proposals that will have a functional effect on the entire mill. For example, the effect of enlarging the capacity of a given operating unit is shown in terms of processing flows throughout the mill, including bottlenecks created or eliminated. The effect of changes in processing-locations or materials-handling facilities may be recognized in terms of delays and transit times. The effect of proposed changes in facility maintenance or the substitution of more reliable equipment can be determined by examining the mill's overall performance under various assumptions as to breakdown frequency at given processing units.

Management controls

Budgets, time/cost standards, accounting methods, and other forms of controls imposed from the tactical management level can be tested. Data showing performance in both physical production and accounting terms can be used to see how accurately given accounting methods reflect processing performance. For example, the conventional cost-accounting system now incorporated in the model could be replaced by a direct-costing system to determine which provides the more useful measure of operations.

Policy formulation

The model provides a way of testing a broad range of policies formulated at the tactical-management level and higher. For example, proposed constraints on in-plant inventory can be evaluated by comparing reductions in working-capital requirements against facility utilization and delivery performance. Whether the mill should operate solely on a produce-to-order basis or maintain finished inventories of common grades of products can be studied by looking at the effects the alternative policies have on operating costs and delivery performance.

Other policy questions that can be explored concern limits on production rates and product mix, the backlog-demand levels that justify overtime or additional shifts, and pricing and quantity-discount schedules. The model might also be useful in determining the effects of changes in work rules and wage levels to provide evaluations related to labor negotiations.

Planning

Finally, the model is also a useful planning tool. Rapid and frequent projections, based on various assumptions as to price, demand, wages, and material costs, can be generated. Or, given a profit objective, the model can be used to determine minimum price-demand combinations required to meet the objective. This information is useful both in determining the feasibility of an objective and in formulating sales and operating plans to achieve the objective.

A major advantage of enterprise models in the planning area is speed. The time and manpower needed to generate forecasts and projections manually limits most planning efforts to a choice between a few selected alternatives. The same factors also limit the number of times a plan can be re-examined and revised. The speed with which enterprise models generate forecasts and projections provides management with the ability to evaluate a much larger range of alternatives.

Summary and Conclusions

The main advantages of enterprise models lie in three of their functional characteristics.

First of all, they are applicable to a broader range of management activities than more specialized models. They can be applied to various levels of the vertical management hierarchy and are useful across a spectrum of staff activities performed at that level. For example, in contrast to the steel-mill model just described, the model of the aerospace firm, mentioned at the beginning of this article, reflects the viewpoint of tactical and strategic management as depicted in Figure 2. The bulk of the model is comprised of decision rules for performing the functions of these two levels. The following list of some of the applications of this model indicate how it parallels the tactical and strategic staff functions:

Evaluation of methods for forecasting overall manpower requirements.

Development of economic criteria for manpower smoothing.

Formulation of bidding policies.

Formulation of R & D policies and programs.

Evaluation of long-range financing policies.

Projections of working capital requirements under various government progress-payment policies.

Evaluation of leased versus owned plant facilities.

Evaluation of profit goals.

Formulation of financial and operating plans to meet a specified profit goal.

Identification of financial constraints on growth rate.

Secondly, an enterprise model can represent management objectives realistically. Real-world business organizations seldom attempt to optimize their operations on the basis of maximizing profits or minimizing costs in the narrow sense. Most formal optimizing techniques make such assumptions. Management in the real world is more apt to seek a currently feasible balance between a complex set of shifting goals and pressures. While no model can depict such a problem in all of its detail, enterprise models do provide a more comprehensive view of the real world than other techniques.

Finally, the enterprise model has great advantages in terms of model-user communication. Managers can understand flow charts more easily than equations and model outputs can be provided in printouts whose format is familiar to the user.

References

BOYD, D. F., and KRASNOW, H. S. (1963), 'Economic evaluation of management information systems', *I.B.M. Systems J.*, vol. 2.

CYERT, R. M., and MARCH, J. G. (1963), *A Behavioral Theory of the Firm*, Prentice-Hall.

FORRESTER, J. W. (1961), *Industrial Dynamics*, M.I.T. Press and Wiley.

GORDON, G. (1962), 'A general purpose systems simulator', *I.B.M. Systems J.*, vol. 1.

MARKOWITZ, H., HAUSNER, B., and KARR, H. (1963), *Simscript: A Simulation Programming Language*, Prentice-Hall.

Further Reading

M. ASIMOW, *Introduction to Design*, Prentice-Hall, 1962.

E. H. BOWMAN and R. B. FETTER, *Analysis for Production and Operations Management*, 3rd edn, Richard D. Irwin, 1961.

J. R. BRIGHT (ed.), *Technological Forecasting for Industry and Government: Methods and Applications*, Prentice-Hall, 1968.

R. G. BROWN, *Statistical Forecasting for Inventory Control*, McGraw-Hill, 1959.

E. S. BUFFA, *Readings in Production and Operations Management*, Wiley, 1966.

B. V. DEAN (ed.), *Operations Research in Research and Development*, Wiley, 1963.

D. P. ECKMAN (ed.), *Systems: Research and Design Proceedings of the First Systems Symposium at Case Institute of Technology*, Wiley, 1961.

S. E. ELMAGHRABY, *The Design of Production Systems*, Reinhold, 1966.

L. J. GARRETT and M. SILVER, *Production Management Analysis*, Harcourt, Brace & World, 1966.

J. W. GAVETT, *Production and Operations Management*, Harcourt, Brace & World, 1968.

W. J. J. GORDON, *Synectics: The Development of Creative Capacity*, Harper & Row, 1961.

V. C. HARE, Jr, *Systems Analysis: A Diagnostic Approach*, Harcourt, Brace & World, 1967.

A. KAUFMANN, *Methods and Models of Operations Research*, Prentice-Hall, 1963.

A. M. LEE, *Applied Queueing Theory*, Macmillan, 1966.

J. F. MAGEE and D. M. BOODMAN, *Production Planning and Inventory Control*, 2nd edn, McGraw-Hill, 1967.

C. MCMILLAN and R. F. GONZALES, *Systems Analysis*, Richard D. Irwin, 1968.

D. W. MILLER and M. K. STARR, *Executive Decisions and Operations Research*, 2nd edn, Prentice-Hall, 1969.

T. L. SAATY, *Elements of Queueing Theory*, McGraw-Hill, 1961.

D. A. SCHON, *Technology and Change: The New Heraclitus*, Delacorte Press, 1967.

M. K. STARR and D. W. MILLER, *Inventory Control: Theory and Practice*, Prentice-Hall, 1962.

M. K. STARR, *Decision Theory and Product Design*, Prentice-Hall, 1963.

M. K. STARR, *Production Management: Systems and Synthesis*, Prentice-Hall, 1964.

I. G. WILSON and M. E. WILSON, *Information, Computers and Systems Design*, Wiley, 1965.

Acknowledgements

Permission to reproduce the Readings in this volume is acknowledged from the following sources:

Reading 1 Prentice-Hall Inc.
Reading 2 *Harvard Business Review*
Reading 3 Harper & Row Publishing Company
Reading 4 Delacorte Press and Pergamon Press Ltd
Reading 5 John Wiley & Sons Inc.
Reading 6 *Harvard Business Review*
Reading 7 *Industrial Management Review*
Reading 8 McGraw-Hill Book Company
Reading 9 The Macmillan Company
Reading 10 The Institute of Management Sciences
Reading 11 The American Institute of Industrial Engineers Inc.
Reading 12 Human Factors Society, L. E. Freund and T. L. Sadosky
Reading 13 Harcourt, Brace & World Inc.
Reading 14 Richard D. Irwin Inc. and the Operations Research Society of America
Reading 15 Prentice-Hall Inc.
Reading 16 Prentice-Hall Inc.
Reading 17 McGraw-Hill Book Company
Reading 18 John Wiley & Sons Inc.
Reading 19 John Wiley & Sons Inc.
Reading 20 Reinhold Publishing Corporation and Litton Educational Publishing Inc.
Reading 21 *Industrial Management Review*

Author Index

Subject Index